THE ORIGINS OF MODERN
PHILOSOPHY OF SCIENCE
1830–1914

Printed and Bound by
Antony Rowe Ltd., Chippenham, Wiltshire

POPULAR LECTURES ON SCIENTIFIC SUBJECTS

Volume II

Hermann von Helmholtz

ROUTLEDGE/THOEMMES PRESS

This edition published by Routledge/Thoemmes Press, 1996

Routledge/Thoemmes Press
11 New Fetter Lane
London EC4P 4EE

The Origins of Modern Philosophy of Science 1830–1914
12 Volumes : ISBN 0 415 13267 3

This is a reprint of the 1895 edition

Routledge / Thoemmes Press is a joint imprint
of Routledge and Thoemmes Antiquarian Books Ltd.

British Library Cataloguing-in-Publication Data
A CIP record of this set is available from the British Library

Publisher's Note

These reprints are taken from original copies of each book. In many cases the condition of those originals is not perfect, the paper, often handmade, having suffered over time and the copy from such things as inconsistent printing pressures resulting in faint text, show-through from one side of a leaf to the other, the filling in of some characters, and the break up of type. The publisher has gone to great lengths to ensure the quality of these reprints but points out that certain characteristics of the original copies will, of necessity, be apparent in reprints thereof.

POPULAR LECTURES

ON

SCIENTIFIC SUBJECTS

BY

HERMANN VON HELMHOLTZ

TRANSLATED BY

E. ATKINSON, Ph.D., F.C.S.

FORMERLY PROFESSOR OF EXPERIMENTAL SCIENCE, STAFF COLLEGE

SECOND SERIES

WITH AN AUTOBIOGRAPHY OF THE AUTHOR

New Edition

LONDON
LONGMANS, GREEN, AND CO.
AND NEW YORK
1895

All rights reserved

PREFACE.

THE FAVOUR with which the first series of Professor Helmholtz's Lectures has been received would justify, if a justification were needed, the publication of the present volume.

I have to express my acknowledgments to Professor G. Croome Robertson, the editor, and to Messrs. Macmillan, the publishers of 'Mind,' for permission to use a translation of the paper on the 'Axioms of Modern Geometry' which appeared in that journal.

The article on 'Academic Freedom in German Universities' contains some statements respecting the Universities of Oxford and Cambridge to which exception has been taken. These statements were a fair representation of the impression produced on the mind of a foreigner by a state of things which no longer exists in those Universities, at least to the same extent. The reform in the University system, which may be said to date from the year 1854, has brought about so many alterations both in the form and in the spirit of the regulations, that older members of the University have been known to speak of the place as so changed that they could scarcely recognise it. Hence, in respect of this article, I have availed myself of the liberty granted by Professor Helmholtz, and have altogether omitted some passages, and have slightly modified others, which would convey an erroneous impression of the present state of things. I have also on these points consulted members of the University on whose judgment I think I can rely.

In other articles, where the matter is of prime importance, I have been anxious faithfully to reproduce the original; nor have I in any such cases allowed a regard for form to interfere with the plain duty of exactly rendering the author's meaning.

E. ATKINSON.

PORTESBERY HILL, CAMBERLEY:
Dec. 1880.

CONTENTS.

LECTURE	PAGE
I. GUSTAV MAGNUS. IN MEMORIAM	1
II. ON THE ORIGIN AND SIGNIFICANCE OF GEOMETRICAL AXIOMS	27
III. ON THE RELATION OF OPTICS TO PAINTING . .	73
I. Form	78
II. Shade	94
III. Colour	110
IV. Harmony of Colour	124
IV. ON THE ORIGIN OF THE PLANETARY SYSTEM . .	139
V. ON THOUGHT IN MEDICINE	199
VI. ON ACADEMIC FREEDOM IN GERMAN UNIVERSITIES .	237
VII. HERMANN VON HELMHOLTZ. AN AUTOBIOGRAPHICAL SKETCH	266

GUSTAV MAGNUS.

In Memoriam.

Address delivered in the Leibnitz Meeting of the Academy of Sciences on July 6, 1871.

THE honourable duty has fallen on me of expressing in the name of this Academy what it has lost in Gustav Magnus, who belonged to it for thirty years. As a grateful pupil, as a friend, and finally as his successor, it was a pleasure to me as well as a duty to fulfil such a task. But I find the best part of my work already done by our colleague Hofmann at the request of the German Chemical Society, of which he is the President. He has solved the difficulty of giving a picture of the life and work of Magnus in the most complete and most charming manner. He has not only anticipated me, but he stood in much closer and more intimate personal relation to Magnus than I did; and, on the other hand, he is much better qualified than I to pronounce a competent judgment on the principal side of Magnus's activity, namely, the chemical.

Hence what remains for me to do is greatly restricted. I shall scarcely venture to speak as the biographer of Magnus, but only of what he was to us and to science, to represent which is our allotted task.

His life was not indeed rich in external events and changes; it was the peaceful life of a man who, free from the cares of outer circumstances, first as member, then as leader of an esteemed, gifted, and amiable family, sought and found abundant satisfaction in scientific work, in the utilisation of scientific results for the instruction and advantage of mankind. Heinrich Gustav Magnus was born in Berlin on May 2, 1802, the fourth of six brothers, who by their talents have distinguished themselves in various directions. The father, Johann Matthias, was chief of a wealthy commercial house, whose first concern was to secure to his children a free development of their individual capacity and inclinations. Our departed friend showed very early a greater inclination for the study of mathematics and natural philosophy than for that of languages. His father arranged his instruction accordingly, by removing him from the Werder Gymnasium and sending him to the Cauer Private Institute, in which more attention was paid to scientific subjects.

From 1822 to 1827 Magnus devoted himself entirely to the study of natural science at the University of Berlin. Before carrying out his original intention of qualifying as a professor of technology, he spent two years with that object in travelling; he remained with Berzelius a long time in Stockholm, then with Dulong, Thénard and Gay-Lussac in Paris. Unusually well prepared by these means, he qualified in the University of this place in technology, and afterwards also in physics; he was appointed extraordinary professor in 1834, and ordinary professor in 1845, and so distinguished himself by his scientific labours at this time, that nine years after his habilitation, on January 27, 1840, he was elected a member of this Academy. From 1832 until 1840 he taught physics in the Artillery and Engineering School; and from 1850 until 1856 chemical technology in the Gewerbe Institut. For a long time he gave the lectures in his own house, using his own instruments, which gradually developed into the most splendid physical collection in existence at that time, and which the State afterwards purchased for the University. His lectures were afterwards given in the University, and he only retained the laboratory in his own house for his own private work and for that of his pupils.

His life was passed thus in quiet but unremitting activity; travels, sometimes for scientific or technical studies, sometimes also in the service of the State, and occasionally for recreation, interrupted his work here from time to time. His experience and knowledge of business were often in demand by the State on various commissions; among these may be especially mentioned the part he took in the chemical deliberations of the Agricultural Board (*Landes-Economie Collegium*), to which he devoted much of his time; above all to the great practical questions of agricultural chemistry.

After sixty-seven years of almost undisturbed health he was overtaken by a painful illness towards the end of the year 1869.[1] He still continued his lectures on physics until February 25, 1870, but during March he was scarcely able to leave his bed, and he died on April 4.

Magnus's was a richly endowed nature, which under happy external circumstances could develop in its own peculiar manner, and was free to choose its activity according to its own mind. But this mind was so governed by reason, and so filled, I might almost say, with artistic harmony, which shunned the immoderate and impure, that he knew how to choose the object of his work wisely, and on this account almost always to attain it. Thus the direction and manner of Magnus's activity accorded so perfectly with his intellectual nature as is the case only with the happy few among

[1] Carcinoma recti.

mortals. The harmonious tendency and cultivation of his mind could be recognised in the natural grace of his behaviour, in the cheerfulness and firmness of his disposition, in the warm amiability of his intercourse with others. There was in all this, much more than the mere acquisition of the outer forms of politeness can ever reach, where they are not illuminated by a warm sympathy and by a fine feeling for the beautiful.

Accustomed from an early age to the regulated and prudent activity of the commercial house in which he grew up, he retained that skill in business which he had so frequently to exercise in the administration of the affairs of this Academy, of the philosophical faculty, and of the various Government commissions. He retained from thence the love of order, the tendency towards the actual, and towards what is practically attainable, even although the chief aim of his activity was an ideal one. He understood that the pleasant enjoyment of an existence free from care, and intercourse with the most amiable circle of relatives and friends, do not bring a lasting satisfaction; but work only, and unselfish work for an ideal aim. Thus he laboured, not for the increase of riches, but for science; not as a dilettante and capriciously, but according to a fixed aim and indefatigably; not in vanity, catching at striking discoveries, which might at once have made his name celebrated. He was, on the contrary, a master of faithful, patient, and modest work, who tests that work again and again, and never ceases until he knows there is nothing left to be improved. But it is also such work, which by the classical perfection of its methods, by the accuracy and certainty of its results, merits and gains the best and most enduring fame. There are among the labours of Magnus masterpieces of finished perfection, especially those on the expansion of gases by heat, and on the tension of vapours. Another master in this field, and one of the most experienced and distinguished, namely, Regnault of Paris, worked at these subjects at the same time with Magnus, but without knowing of his researches. The results of both investigators were made public almost simultaneously, and showed by their extraordinarily close agreement with what fidelity and with what skill both had laboured. But where differences showed themselves, they were eventually decided in favour of Magnus.

The unselfishness with which Magnus held to the ideal aim of his efforts is shown in quite a characteristic manner, in the way in which he attracted younger men to scientific work, and as soon as he believed he had discovered in them zeal and talent for such work by placing at their disposal his apparatus, and the appliances of his private laboratory. This was the way in which I was brought in close relation to him, when I found myself in Berlin for the purpose of passing the Government medical examination.

He invited me at that time (I myself would not have ventured to propose it) to extend my experiments on fermentation and putrefaction in new directions, and to apply other methods, which required greater means than a young army surgeon living on his pay could provide himself with. At that time I worked with him almost daily for about three months, and thus gained a deep and lasting impression of his goodness, his unselfishness, and his perfect freedom from scientific jealousy.

By such a course he not only surrendered the external advantages which the possession of one of the richest collections of instruments would have secured an ambitious man against competitors, but he also bore with perfect composure the little troubles and vexations involved in the want of skill and the hastiness with which young experimentalists are apt to handle costly instruments. Still less could it be said that, after the manner of the learned in other countries, he utilised the work of the pupils for his own objects, and for the glorification of his own name. At that time chemical laboratories were being established according to Liebig's precedent: of physical laboratories—which, it may be observed, are much more difficult to organise—not one existed at that time to my knowledge. In fact, their institution is due to Magnus.

In such circumstances we see an essential part of the inner tendency of the man, which must not be neglected in estimating his value: he was not only an investigator, he was also a teacher of science in the highest and widest sense of the word. He did not wish science to be confined to the study and lecture-room, he desired that it should find its way into all conditions of life. In his active interest for technology, in his zealous participation in the work of the Agricultural Board, this phase of his efforts was plainly reflected, as well as in the great trouble he took in the preparation of experiments, and in the ingenious contrivance of the apparatus required for them.

His collection of instruments, which subsequently passed into the possession of the University, and is at present used by me as his successor, is the most eloquent testimony of this Everything is in the most perfect order: if a silk-thread, a glass tube, or a cork, are required for an experiment, one may safely depend on finding them near the instrument. All the apparatus which he contrived is made with the best means at his disposal, without sparing either material, or the labour of the workman, so as to ensure the success of the experiment, and by making it on a sufficiently large scale to render it visible as far off as possible. I recollect very well with what wonder and admiration we students saw him experiment, not merely because all the experiments were successful and brilliant, but because they scarcely seemed to occupy or to disturb his thoughts. The easy and clear flow of his discourse went on without interruption; each experiment came in its right place, was performed quickly, without haste or hesitation, and was then put aside.

I have already mentioned that the valuable collection of apparatus came into the possession of the University during his lifetime. He specially wished that what he had collected and constructed as appliances in his scientific work should not be scattered and estranged from the original purpose to which he had devoted his life. With this feeling he bequeathed to the University the rest of the apparatus of his laboratory, as well as his very rich and valuable library, and he thus laid the foundation for the further development of a Public Physical Institute.

It is sufficient in these few touches to have recalled the mental individuality of our departed friend, so far as the sources of the direction of his activity are to be found.

Personal recollections will furnish a livelier image to all those of you who have worked with him for the last thirty years.

If we now proceed to discuss the results of his researches it will not be sufficient to read through and to estimate his academical writings. I have already shown that a prominent part of his activity was directed to his fellow-creatures. To this must be added, that he lived in an age when natural science passed through a process of development, with a rapidity such as never occurred before in the history of science. But the men who belonged to such a time, and co-operated in this development are apt to appear in wrong perspective to their successors, since the best part of their work seems to the latter self-evident, and scarcely worthy of mention.

It is difficult for us to realise the condition of natural science as it existed in Germany, at least in the first twenty years of this century. Magnus was born in 1802; I myself nineteen years later; but when I go back to my earliest recollections, when I began to study physics out of the school-books in my father's possession, who was himself taught in the Cauer Institute, I still see before me the dark image of a series of ideas which seems now like the alchemy of the middle ages. Of Lavoisier's and of Humphry Davy's revolutionising discoveries, not much had got into the school-books. Although oxygen was already known, yet phlogiston, the fire element, played also its part. Chlorine was still oxygenated hydrochloric acid; potash and lime were still elements. Invertebrate animals were divided into insects and reptiles; and in botany we still counted stamens.

It is strange to see how late and with what hesitation Germans applied themselves to the study of natural science in this century, whilst they had taken so prominent a part in its earlier development. I need only name Copernicus, Kepler, Leibnitz, and Stahl.

For we may indeed boast of our eager, fearless and unselfish love of truth, which flinches before no authority, and is stopped by no pretence; shuns no sacrifice and no labour, and is very modest in its claims on worldly success. But even on this account she ever impels us first of all to pursue the questions of principle to their ultimate sources, and to trouble ourselves but little about what has no connection with fundamental principles, and especially about practical consequences and about useful applications. To this must be added another reason, namely, that the independent mental development of the last three hundred years, began under political conditions which caused the chief weight to fall on theological studies. Germany has liberated Europe from the tyranny of the ancient church; but she has also paid a much dearer price for this freedom than other nations. After the religious wars, she remained devastated, impoverished, politically shattered, her boundaries spoiled, and arrogantly handed over defenceless to her neighbours. To deduce consequences from the new moral views, to prove them scientifically, to work them out in all regions of intellectual life, for all this there was no time during the storm of war; each man had to hold to his own party, every incipient change of opinion was looked upon as treachery, and excited bitter wrath. Owing to the Reformation, intellectual life had lost its old stability and cohesion; everything appeared in a new light, and new questions arose. The German mind could not be quieted with outward uniformity; when it was not convinced and satisfied, it did not allow its doubt to remain silent. Thus it was theology, and next to it classical philology and philosophy, which, partly as scientific aids of theology, partly for what they could do for the solution of the new moral, æsthetical, and metaphysical problems, laid claim almost exclusively to the interest of scientific culture. Hence it is clear why the Protestant nations, as well as that part of the Catholics which, wavering in its old faith, only remained outwardly in connection with its church, threw itself with such zeal on philosophy. Ethical and metaphysical problems were chiefly to be solved; the sources of knowledge had to be critically examined, and this was done with deeper earnestness than formerly. I need not enumerate the actual results which the last century gained by this work. It excited soaring hopes, and it cannot be denied that metaphysics has a dangerous attraction for the German mind; it could not again abandon it until all its hiding-places had been searched through and it had satisfied itself that for the present nothing more is to be found there.

Then, in the second half of the last century, the rejuvenescent intellectual life of the nation began to cultivate its artistic flowers; the clumsy language transformed itself into one of the most expressive instruments of the human mind; out of what was still the hard, poor, and wearisome condition of civil and political life, the results of the religious war, in which the figure of the Prussian hero-king only now cast the first hope of a better future, to be again followed by the misery of the Napoleonic war,—out of this joyless existence, all sensitive minds gladly fled into the flowery land opened out by German poetry, rivalling as it did the best poetry of all times and of all peoples; or in the sublime aspects of philosophy they endeavoured to sink reality in oblivion.

And the natural sciences were on the side of this real world, so willingly overlooked. Astronomy alone could at that time offer great and sublime prospects; in all other branches long and patient labour was still necessary before great principles could be attained; before these subjects could have a voice in the great questions of human life; or before they became the powerful means of the authority of man over the forces of nature which they have since become. The labour of the natural philosopher seems narrow, low, and insignificant compared with the great conceptions of the philosopher and of the poet; it was only those natural philosophers who, like Oken, rejoiced in poetical philosophical conceptions, who found willing auditors.

Far be it from me as a one-sided advocate of scientific interests to blame this period of enthusiastic excitement; we have, in fact, to thank it for the moral force which broke the Napoleonic yoke; we have to thank it for the grand poetry which is the noblest treasure of our nation; but the real world retains its right against every semblance, even against the most beautiful; and individuals, as well as nations, who wish to rise to the ripeness of manhood must learn to look reality in the face, in order to bend it to the purpose of the mind. To flee into an ideal world is a false resource of transient success; it only facilitates the play of the adversary; and when knowledge only reflects itself, it becomes unsubstantial and empty, or resolves itself into illusions and phrases.

Against the errors of a mental tendency, which corresponded at first to the natural soaring of a fresh youthful ambition, but which afterwards, in the age of the Epigones of the Romantic school and of the philosophy of Identity, fell into sentimental straining after sublimity and inspiration, a reaction took place, and was carried out not merely in the regions of science, but also in history, in art, and in philology. In the last departments, too, where we deal directly with products of activity of the human mind, and where, therefore, a construction *à priori* from the psychological laws seems much more possible than in nature, it has come to be understood that we must first know the facts, before we can establish their laws.

Gustav Magnus's development happened during the period of this struggle; it lay in the whole tendency of his mind, that he whose gentle spirit usually endeavoured to reconcile antagonisms, took a decided part in favour of pure experience as against speculation. If he forbore to wound people, it must be confessed that he did not relax one iota of the principle which, with sure instinct, he had recognised as the true one; and in the most influential quarters he fought in a twofold sense; on the one hand, because in physics it was a question as to the foundations of the whole of natural sciences; and on the other hand, because the University of Berlin, with its numerous students, had long been the stronghold of speculation. He continually preached to his pupils that no reasoning, however plausible it might seem, avails against actual fact, and that observation and experiment must decide; and he was always anxious that every practicable experiment should be made which could give practical confirmation or refutation of an assumed law. He did not limit in any way the applicability of scientific methods in the investigation of inanimate nature, but in his research on the gases of the blood (1837) he dealt a blow at the heart of vitalistic theories. He led physics to the centre of organic change, by laying a scientific foundation for a correct theory of respiration; a foundation upon which a great number of more recent investigators have built, and which has developed into one of the most important chapters of physiology.

He cannot be reproached with having had too *little* confidence in carrying out his principle; but I must confess that I myself and many of my companions formerly thought that Magnus carried his distrust of speculation too far, especially in relation to mathematical physics. He had probably never dipped very deep in the latter subject, and that strengthened our doubts. Yet when we look around us from the standpoint which science has now attained, it must be confessed that his distrust of the mathematical physics of that date was not unfounded. At that time no separation had been distinctly made as to what was empirical matter of fact, what mere verbal definition, and what only hypothesis. The vague mixture of these elements which formed the basis of calculation was put forth as axioms of metaphysical necessity, and postulated a similar kind of necessity for the results. I need only recall to you the great part which hypotheses as to the atomic structure of bodies played in mathematical physics during the first half of this century, whilst as good as nothing was known of atoms; and, for instance, hardly anything was known of the extraordinary influence which heat has on molecular forces. We now know that the expansive force of gases depends on motion due to heat; at that period most physicists regarded heat as imponderable matter.

In reference to atoms in molecular physics, Sir W. Thomson says, with much weight, that their assumption can explain no property of the body which has not previously been attributed to the atoms. Whilst assenting to this opinion, I would in no way express myself against the existence of atoms, but only against the endeavour to deduce the principles of theoretical physics from purely hypothetical assumptions as to the atomic structure of bodies. We now know that many of these hypotheses, which found favour in their day, far overshot the mark. Mathematical physics has acquired an entirely different character under the hands of Gauss, of F. E. Neumann and their pupils, among the Germans; as well as from those mathematicians who in England followed Faraday's lead, Stokes, W. Thomson, and Clerk-Maxwell. It is now understood that mathematical physics is a purely experimental science; that it has no other principles to follow than those of experimental physics. In our immediate experience we find bodies variously formed and constituted; only with such can we make our observations and experiments. Their actions are made up of the actions which each of their parts contributes to the sum of the whole; and hence, if we wish to know the simplest and most general law of the action of masses and substances found in nature upon one another, and if we wish to divest these laws of the accidents of form, magnitude, and position of the bodies concerned, we must go back to the laws of action of the smallest particles, or, as mathematicians designate it, the elementary volume. But these are not, like the atoms, disparate and heterogeneous, but continuous and homogeneous.

The characteristic properties of the elementary volumes of different bodies are to be found experimentally, either directly, where the knowledge of the sum is sufficient to discover the constituents, or hypothetically, where the calculated sum of effects in the greatest possible number of different cases must be compared with actual fact by observation and by experiment. It is thus admitted that mathematical physics only investigates the laws of action of the elements of a body independently of the accidents of form, in a purely empirical manner, and is therefore just as much under the control of experience as what are called experimental physics. In principle they are not at all different, and the former only continues the function of the latter, in order to arrive at still simpler and still more general laws of phenomena.

It cannot be doubted that this analytical tendency of physical inquiry has assumed another character; that it has just cast off that which was the means of placing Magnus towards it in some degree of antagonism. He tried to maintain, at least in former years, that the business of the mathematical and that of the experimental physicist are quite distinct from one another; that a young man who wishes to pursue physics would have to decide between the two. It appears to me, on the contrary, that the conviction is constantly gaining ground, that in the present more advanced state of science those only can experimentalise profitably who have a clear-sighted knowledge of theory, and know how to propound and pursue the right questions; and, on the other hand, only those can theorise with advantage who have great practice in experiments. The discovery of spectrum analysis is the most brilliant example within our recollection of such an interpenetration of theoretical knowledge and experimental skill.

I am not aware whether Magnus subsequently expressed other views as to the relation of experimental and mathematical physics. In any case, those who regard his former desertion of mathematical physics as a reaction against the misuse of speculation carried too far, must also admit that in the older mathematical physics there are many reasons for this dislike, and that, on the other hand, he received with the greatest pleasure the results which Kirchhoff, Sir W. Thomson, and others had developed out of new facts from theoretical starting-points. I may here be permitted to adduce my own experience. My researches were mostly developed in a manner against which Magnus tried to guard; yet I never found in him any but the most willing and friendly recognition.

It is, however, natural that every one, relying upon his own experience, should recommend to others, as most beneficial, the way which best suits his own nature, and by which he has made the quickest progress. And if we are all of the same opinion that the task of science is to find the *Laws of Facts*, then each one may be left free either to plunge into facts, and to search where he might come upon traces of laws still unknown, or from laws already known to search out the points where new facts are to be discovered. But just as we all, like Magnus, are opposed to the theorist who holds it unnecessary to prove experimentally the hypothetical results which seem axioms to him, so would Magnus—as his works decidedly show—pronounce with us against that kind of excessive empiricism which sets out to discover facts which fit to no rule, and which also try carefully to avoid a law, or a possible connection between newly discovered facts.

It must here be mentioned that Faraday, another great physicist, worked in England exactly in the same direction, and with the same object; to whom, on that account, Magnus was bound by the heartiest sympathy. With Faraday, the antagonism to the physical theories hitherto held, which treated of atoms and forces acting at a distance, was even more pronounced than with Magnus.

We must, moreover, admit that Magnus mostly worked with success on problems which seemed specially adapted to mathematical treatment; as, for instance, his research on the deviation of rotating shot fired from rifled guns; also his paper on the form of jets of water and their resolution into drops. In the first, he proved, by a very cleverly arranged experiment, how the resistance of the air, acting on the ball from below, must deflect it sideways as a rotating body, in a direction depending on that of the rotation; and how, in consequence of this, the trajectory is deflected in the same direction. In the second treatise, he investigated the different forms of jets of water, how they are partly changed by the form of the aperture through which they flow, partly in consequence of the manner in which they flow to it; and how their resolution into drops is caused by external agitation.

He applied the principle of the stroboscopic disk in observing the phenomena, by looking at the jet through small slits in a rotating disk. He grouped the various phenomena with peculiar tact, so that those among them which are alike were easily seen, and one elucidated the other. And if a final mechanical explanation is not always attained, yet the reason for a great number of characteristic features of the individual phenomena is plain. In this respect many of his researches—I might specially commend those on the efflux of jets of water—are excellent models of what Goethe theoretically advanced, and in his physical labours endeavoured to accomplish, though with only partial success.

But even where Magnus from his standpoint, and armed with the knowledge of his time, exerted himself in vain to seize the kernel of the solution of a difficult question, a host of new and valuable facts is always brought to light. Thus in his research on the thermoelectric battery, where he correctly saw that a critical question was to be solved, and at the conclusion declared: 'When I commenced the experiment just described, I confidently hoped to find that thermoelectrical currents are due to a motion of heat.' In this sense he investigated the cases in which the thermo-electrical circuit consisted of a single metal in which there were alternately hard portions, and such as had been softened by heat; or those in which the parts in contact had very different temperatures. He was convinced that the thermo-electrical current was due neither to the radiating power, nor to the conductivity for heat, using this expression in its ordinary meaning, and he had to content himself with the obviously imperfect explanation that two pieces of the same metal at different temperatures acted like dissimilar conductors, which like liquids do not fall in with the potential series. The solution was first furnished by the two general laws of the mechanical theory of heat. Magnus's hope was not unfulfilled. Sir W. Thomson discovered that alterations in the conductivity for heat, though such as were produced by the electrical current itself, were indeed the sources of the current.

It is the nature of the scientific direction which Magnus pursued in his researches, that they build many a stone into the great fabric of science, which give it an ever broader support, and an ever growing height, without its appearing to a fresh observer as a special and distinctive work due to the sole exertion of any one scientific man. If we wish to explain the importance of each stone for its special place, how difficult to procure it, and how skilfully it was worked, we must presuppose either that the hearer knows the entire history of the building, or we must explain it to him, by which more time is lost than I can now claim.

Thus it is with Magnus's researches. Wherever he has attacked, he has brought out a host of new and often remarkable facts; he has carefully and accurately observed them, and he has brought them in connection with the great fabric of science. He has, moreover, bequeathed to science a great number of ingenious and carefully devised new methods, as instruments with which future generations will continue to discover hidden veins of the noble metal of everlasting laws in the apparently waste and wild chaos of accident. Magnus's name will always be mentioned in the first line of those on whose labours the proud edifice of the science of Nature reposes; of the science which has so thoroughly remodelled the life of modern humanity by its intellectual influence, as well as by its having subjugated the forces of nature to the dominion of the mind.

I have only spoken of Magnus's physical labours, and have only mentioned those which seemed to me characteristic for his individuality. But the number of his researches is very great, and they extend over wider regions than could now be grasped by any single inquirer. He began as a chemist, but even then he inclined to those cases which showed remarkable physical conditions; he was afterwards exclusively a physicist. But parallel with this he cultivated a very extended study of technology, which of itself would alone have occupied a man's life.

He has departed, after a rich life and a fruitful activity. The old law that no man's life is free from pain must have been applied to him also; and yet his life seems to have been especially happy. He had what men generally desire most; but he knew how to ennoble external fortune by putting it at the service of unselfish objects. To him was granted, what is dearest to the mind of a noble spirit, to dwell in the centre of an affectionate family, and in a circle of faithful and distinguished friends. But I count his rarest happiness to be that he could work in pure enthusiasm for an ideal principle; and that he saw the cause which he served go on victoriously, and develop to unheard of wealth and ever wider activity.

And in conclusion we must add, in so far as thoughtfulness, purity of intention, moral and intellectual tact, modesty, and true humanity can rule over the caprices of fortune and of man, in so far was Magnus the artificer of his own fortune; one of the most satisfactory and contented natures, who secure the love and favour of men, who with sure inspiration know how to find the right place for their activity; and of whom we may say, envious fact does not embitter their successes, for, working for pure objects and with pure wishes, they would find contentment even without external successes.

ON THE

ORIGIN AND SIGNIFICANCE

OF

GEOMETRICAL AXIOMS.

Lecture delivered in the Docenten Verein in Heidelberg, in the year 1870.

THE fact that a science can exist and can be developed as has been the case with geometry, has always attracted the closest attention among those who are interested in questions relating to the bases of the theory of cognition. Of all branches of human knowledge, there is none which, like it, has sprung as a completely armed Minerva from the head of Jupiter; none before whose death-dealing Aegis doubt and inconsistency have so little dared to raise their eyes. It escapes the tedious and troublesome task of collecting experimental facts, which is the province of the natural sciences in the strict sense of the word; the sole form of its scientific method is deduction. Conclusion is deduced from conclusion, and yet no one of common sense doubts but that these geometrical principles must find their practical application in the real world about us. Land surveying, as well as architecture, the construction of machinery no less than mathematical physics, are continually calculating relations of space of the most varied kind by geometrical principles; they expect that the success of their constructions and experiments shall agree with these calculations; and no case is known in which this expectation has been falsified, provided the calculations were made correctly and with sufficient data.

Indeed, the fact that geometry exists, and is capable of all this, has always been used as a prominent example in the discussion on that question, which forms, as it were, the centre of all antitheses of philosophical systems, that there can be a cognition of principles destitute of any bases drawn from experience. In the answer to Kant's celebrated question, 'How are synthetical principles *a priori* possible?' geometrical axioms are certainly those examples which appear to show most decisively that synthetical principles are *a priori* possible at all. The circumstance that such principles exist, and force themselves on our conviction, is regarded as a proof that space is an *a priori* mode of all external perception.

It appears thereby to postulate, for this *a priori* form, not only the character of a purely formal scheme of itself quite unsubstantial, in which any given result experience would fit; but also to include certain peculiarities of the scheme, which bring it about that only a certain content, and one which, as it were, is strictly defined, could occupy it and be apprehended by us.[1]

It is precisely this relation of geometry to the theory of cognition which emboldens me to speak to you on geometrical subjects in an assembly of those who for the most part have limited their mathematical studies to the ordinary instruction in schools. Fortunately, the amount of geometry taught in our gymnasia will enable you to follow, at any rate the tendency, of the principles I am about to discuss.

I intend to give you an account of a series of recent and closely connected mathematical researches which are concerned with the geometrical axioms, their

---

[1] In his book, *On the Limits of Philosophy*, Mr. W. Tobias maintains that axioms of a kind which I formerly enunciated are a misunderstanding of Kant's opinion. But Kant specially adduces the axioms, that the straight line is the shortest (*Kritik der reinen Vernunft*, Introduction, v. 2nd ed. p. 16); that space has three dimensions (*Ibid*. part i. sect. i. § 3, p. 41); that only one straight line is possible between two points (*Ibid*. part ii. sect. i. ' On the Axioms of Intuition'), as axioms which express *a priori* the conditions of intuition by the senses. It is not here the question, whether these axioms were originally given as intuition of space, or whether they are only the starting-points from which the understanding can develop such axioms *a priori* on which my critic insists.

relations to experience, with the question whether it is logically possible to replace them by others.

Seeing that the researches in question are more immediately designed to furnish proofs for experts in a region which, more than almost any other, requires a higher power of abstraction, and that they are virtually inaccessible to the non-mathematician, I will endeavour to explain to such a one the question at issue. I need scarcely remark that my explanation will give no proof of the correctness of the new views. He who seeks this proof must take the trouble to study the original researches.

Anyone who has entered the gates of the first elementary axioms of geometry, that is, the mathematical doctrine of space, finds on his path that unbroken chain of conclusions of which I just spoke, by which the ever more varied and more complicated figures are brought within the domain of law. But even in their first elements certain principles are laid down, with respect to which geometry confesses that she cannot prove them, and can only assume that anyone who understands the essence of these principles will at once admit their correctness. These are the so-called axioms.

For example, the proposition that if the shortest line drawn between two points is called a *straight* line, there can be only one such straight line. Again, it is an axiom that through any three points in space, not lying in a straight line, a plane may be drawn, i.e. a surface which will wholly include every straight line joining any two of its points. Another axiom, about which there has been much discussion, affirms that through a point lying without a straight line only one straight line can be drawn parallel to the first; two straight lines that lie in the same plane and never meet, however far they may be produced, being called parallel. There are also axioms that determine the number of dimensions of space and its surfaces, lines and points, showing how they are continuous; as in the propositions, that a solid is bounded by a surface, a surface by a line and a line by a point, that the point is indivisible, that by the movement of a point a line is described, by that of a line a line or a surface, by that of a surface a surface or a solid, but by the movement of a solid a solid and nothing else is described.

Now what is the origin of such propositions, unquestionably true yet incapable of proof in a science where everything else is reasoned conclusion? Are they inherited from the divine source of our reason as the idealistic philosophers think, or is it only that the ingenuity of mathematicians has hitherto not been penetrating enough to find the proof? Every new votary, coming with fresh zeal to geometry, naturally strives to succeed where all before him have failed. And it is quite right that each should make the trial afresh; for, as the question has hitherto stood, it is only by the fruitlessness of one's own efforts that one can be convinced of the impossibility of finding a proof. Meanwhile solitary inquirers are always from time to time appearing who become so deeply entangled in complicated trains of reasoning that they can no longer discover their mistakes and believe they have solved the problem. The axiom of parallels especially has called forth a great number of seeming demonstrations.

The main difficulty in these inquiries is, and always has been, the readiness with which results of everyday experience become mixed up as apparent necessities of thought with the logical processes, so long as Euclid's method of constructive intuition is exclusively followed in geometry. It is in particular extremely difficult, on this method, to be quite sure that in the steps prescribed for the demonstration we have not involuntarily and unconsciously drawn in some most general results of experience, which the power of executing certain parts of the operation has already taught us practically. In drawing any subsidiary line for the sake of his demonstration, the well-trained geometer always asks if it is possible to draw such a line. It is well known that problems of construction play an essential part in the system of geometry. At first sight, these appear to be practical operations, introduced for the training of learners; but in reality they establish the existence of definite figures. They show that points, straight lines, or circles such as the problem requires to be constructed are possible under all conditions, or they determine any exceptions that there may be. The point on which the investigations turn, that we are about to consider, is essentially of this nature. The foundation of all proof by Euclid's method consists in establishing the congruence of lines, angles, plane figures, solids, &c. To make the congruence evident, the geometrical figures are supposed to be applied to one another, of course without changing their form and dimensions. That this is in fact possible we have all experienced from our earliest youth. But, if we proceed to build necessities of thought upon this assumption of the free translation of fixed figures, with unchanged form, to every part of space, we must see whether the assumption does not involve some presupposition of which no logical proof is given. We shall see later on that it does indeed contain one of the most serious import. But if so, every proof by congruence rests upon a fact which is obtained from experience only.

I offer these remarks, at first only to show what difficulties attend the complete analysis of the presuppositions we make, in employing the common constructive method. We evade them when we apply, to the investigation of principles, the analytical method of modern algebraical geometry. The whole process of algebraical calculation is a purely logical operation; it can yield no relation between the quantities submitted to it that is not already contained in the equations which give occasion for its being applied. The recent investigations in question have accordingly been conducted almost exclusively by means of the purely abstract methods of analytical geometry.

However, after discovering by the abstract method what are the points in question, we shall best get a distinct view of them by taking a region of narrower limits than our own world of space. Let us, as we logically may, suppose reasoning beings of only two dimensions to live and move on the surface of some solid body. We will assume that they have not the power of perceiving anything outside this surface, but that upon it they have perceptions similar to ours. If such beings worked out a geometry, they would of course assign only two dimensions to their space. They would ascertain that a point in moving describes a line, and that a line in moving describes a surface. But they could as little represent to themselves what further spatial construction would be generated by a surface moving out of itself, as we can represent what would be generated by a solid moving out of the space we know. By the much-abused expression 'to represent' or 'to be able to think how something happens' I understand—and I do not see how anything else can be understood by it without loss of all meaning—the power of imagining the whole series of sensible impressions that would be had in such a case. Now as no sensible impression is known relating to such an unheard-of event, as the movement to a fourth dimension would be to us, or as a movement to our third dimension would be to the inhabitants of a surface, such a 'representation' is as impossible as the 'representation' of colours would be to one born blind, if a description of them in general terms could be given to him.

Our surface-beings would also be able to draw shortest lines in their superficial space. These would not necessarily be straight lines in our sense, but what are technically called *geodetic lines* of the surface on which they live; lines such as are described by a *tense* thread laid along the surface, and which can slide upon it freely. I will henceforth speak of such lines as the *straightest* lines of any particular surface or given space, so as to bring out their analogy with the straight line in a plane. I hope by this expression to make the conception more easy for the apprehension of my non-mathematical hearers without giving rise to misconception.

Now if beings of this kind lived on an infinite plane, their geometry would be exactly the same as our planimetry. They would affirm that only one straight line is possible between two points; that through a third point lying without this line only one line can be drawn parallel to it; that the ends of a straight line never meet though it is produced to infinity, and so on. Their space might be infinitely extended, but even if there were limits to their movement and perception, they would be able to represent to themselves a continuation beyond these limits; and thus their space would appear to them infinitely extended, just as ours does to us, although our bodies cannot leave the earth, and our sight only reaches as far as the visible fixed stars.

But intelligent beings of the kind supposed might also live on the surface of a sphere. Their shortest or straightest line between two points would then be an arc of the great circle passing through them. Every great circle, passing through two points, is by these divided into two parts; and if they are unequal, the shorter is certainly the shortest line on the sphere between the two points, but also the other or larger arc of the same great circle is a geodetic or straightest line, *i.e.* every smaller part of it is the shortest line between its ends. Thus the notion of the geodetic or straightest line is not quite identical with that of the shortest line. If the two given points are the ends of a diameter of the sphere, every plane passing through this diameter cuts semicircles, on the surface of the sphere, all of which are shortest lines between the ends; in which case there is an equal number of equal shortest lines between the given points. Accordingly, the axiom of there being only one shortest line between two points would not hold without a certain exception for the dwellers on a sphere.

Of parallel lines the sphere-dwellers would know nothing. They would maintain that any two straightest lines, sufficiently produced, must finally cut not in one only but in two points. The sum of the angles of a triangle would be always greater than two right angles, increasing as the surface of the triangle grew greater. They could thus have no conception of geometrical similarity between greater and smaller figures of the same kind, for with them a greater triangle must have different angles from a smaller one. Their space would be unlimited, but would be found to be finite or at least represented as such.

It is clear, then, that such beings must set up a very different system of geometrical axioms from that of the inhabitants of a plane, or from ours with our space of three dimensions, though the logical powers of all were the same; nor are more examples necessary to show that geometrical axioms must vary according to the kind of space inhabited by beings whose powers of reason are quite in conformity with ours. But let us proceed still farther.

Let us think of reasoning beings existing on the surface of an egg-shaped body. Shortest lines could be drawn between three points of such a surface and a triangle constructed. But if the attempt were made to construct congruent triangles at different parts of the surface, it would be found that two triangles, with three pairs of equal sides, would not have their angles equal. The sum of the angles of a triangle drawn at the sharper pole of the body would depart farther from two right angles than if the triangle were drawn at the blunter pole or at the equator. Hence it appears that not even such a simple figure as a triangle can be moved on such a surface without change of form. It would also be found that if circles of equal radii were constructed at different parts of such a surface (the length of the radii being always measured by shortest lines along the surface) the periphery would be greater at the blunter than at the sharper end.

We see accordingly that, if a surface admits of the figures lying on it being freely moved without change of any of their lines and angles as measured along it, the property is a special one and does not belong to every kind of surface. The condition under which a surface possesses this important property was pointed out by Gauss in his celebrated treatise on the curvature of surfaces.[1] The 'measure of curvature,' as he called it, *i.e.* the reciprocal of the product of the greatest and least radii of curvature, must be everywhere equal over the whole extent of the surface.

Gauss showed at the same time that this measure of curvature is not changed if the surface is bent without distension or contraction of any part of it. Thus we can roll up a flat sheet of paper into the form of a cylinder, or of a cone, without any change in the dimensions of the figures taken along the surface of the sheet. Or the hemispherical fundus of a bladder may be rolled into a spindle-shape without altering the dimensions on the surface. Geometry on a plane will therefore be the same as on a cylindrical surface; only in the latter case we must imagine that any number of layers of this surface, like the layers of a rolled sheet of paper, lie one upon another, and that after each entire revolution round the cylinder a new layer is reached different from the previous ones.

These observations are necessary to give the reader a notion of a kind of surface the geometry of which is on the whole similar to that of the plane, but in which

---

[1] Gauss, *Werke,* Bd. IV. p. 215, first published in *Commentationes Soc. Reg. Scientt. Gottengensis recentiores,* vol. vi., 1828.

the axiom of parallels does not hold good. This is a kind of curved surface which is, as it were, geometrically the counterpart of a sphere, and which has therefore been called the *pseudospherical surface* by the distinguished Italian mathematician E. Beltrami, who has investigated its properties.[1] It is a saddle-shaped surface of which only limited pieces or strips can be connectedly represented in our space, but which may yet be thought of as infinitely continued in all directions, since each piece lying at the limit of the part constructed can be conceived as drawn back to the middle of it and then continued. The piece displaced must in the process change its flexure but not its dimensions, just as happens with a sheet of paper moved about a cone formed out of a plane rolled up. Such a sheet fits the conical surface in every part, but must be more bent near the vertex and cannot be so moved over the vertex as to be at the same time adapted to the existing cone and to its imaginary continuation beyond.

Like the plane and the sphere, pseudospherical surfaces have their measure of curvature constant, so that every piece of them can be exactly applied to every

---

[1] *Saggio di Interpretazione della Geometria Non-Euclidea*, Napoli, 1868.—*Teoria fondamentale degli Spazii di Curvatura costante*, *Annali di Matematica*, Ser. II. Tom. II. pp. 232–55. Both have been translated into French by J. Hoüel, *Annales Scientifiques de l'Ecole Normale*, Tom V., 1869.

other piece, and therefore all figures constructed at one place on the surface can be transferred to any other place with perfect congruity of form, and perfect equality of all dimensions lying in the surface itself. The measure of curvature as laid down by Gauss, which is positive for the sphere and zero for the plane, would have a constant negative value for pseudospherical surfaces, because the two principal curvatures of a saddle-shaped surface have their concavity turned opposite ways.

A strip of a pseudospherical surface may, for example, be represented by the inner surface (turned towards the axis) of a solid anchor-ring. If the plane figure *aabb* (Fig. 1) is made to revolve on its axis of symmetry AB, the two arcs *ab* will describe a pseudospherical concave-convex surface like that of the ring. Above and below, towards *aa* and *bb*, the surface will turn outwards with ever-increasing flexure, till it becomes perpendicular to the axis, and ends at the edge with one curvature infinite. Or, again, half of a pseudospherical surface may be rolled up into the shape of a champagne-glass (Fig. 2), with tapering stem infinitely prolonged. But the surface is always necessarily bounded by a sharp edge beyond which it cannot be directly continued. Only by supposing each single piece of the edge cut loose and drawn along the surface of the ring or glass, can it be brought to places of different flexure, at which farther continuation of the piece is possible.

In this way too the straightest lines of the pseudospherical surface may be infinitely produced. They do not, like those on a sphere, return upon themselves, but, as on a plane, only one shortest line is possible between the two given points. The axiom of parallels does not, however, hold good. If a straightest line is

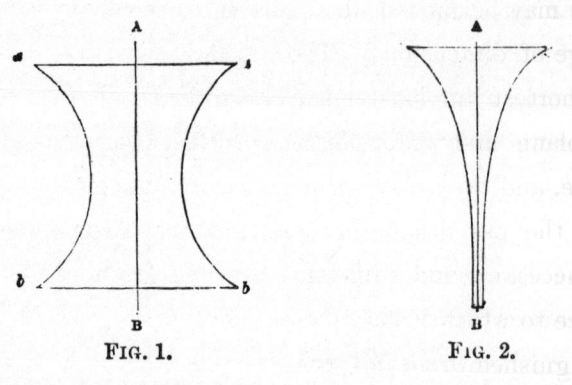

FIG. 1.   FIG. 2.

given on the surface and a point without it, a whole pencil of straightest lines may pass through the point, no one of which, though infinitely produced, cuts the first line; the pencil itself being limited by two straightest lines, one of which intersects one of the ends of the given line at an infinite distance, the other the other end.

Such a system of geometry, which excluded the axiom of parallels, was devised on Euclid's synthetic method, as far back as the year 1829, by N. J. Lobatchewsky, professor of mathematics at Kasan,[1] and it was proved that this system could be carried out as consistently as Euclid's. It agrees exactly with the geometry of the pseudospherical surfaces worked out recently by Beltrami.

Thus we see that in the geometry of two dimensions a surface is marked out as a plane, or a sphere, or a pseudospherical surface, by the assumption that any figure may be moved about in all directions without change of dimensions. The axiom, that there is only one shortest line between any two points, distinguishes the plane and the pseudospherical surface from the sphere, and the axiom of parallels marks off the plane from the pseudosphere. These three axioms are in fact necessary and sufficient, to define as a plane the surface to which Euclid's planimetry has reference, as distinguished from all other modes of space in two dimensions.

The difference between plane and spherical geometry has been long evident, but the meaning of the axiom of parallels could not be understood till Gauss had developed the notion of surfaces flexible without dilatation, and consequently that of the possibly infinite continuation of pseudospherical surfaces. Inhabiting, as we do, a space of three dimensions and endowed with organs of sense for their perception, we

[1] *Principien der Geometrie,* Kasan, 1829-30.

can represent to ourselves the various cases in which beings on a surface might have to develop their perception of space; for we have only to limit our own perceptions to a narrower field. It is easy to think away perceptions that we have; but it is very difficult to imagine perceptions to which there is nothing analogous in our experience. When, therefore, we pass to space of three dimensions, we are stopped in our power of representation, by the structure of our organs and the experiences got through them which correspond only to the space in which we live.

There is however another way of treating geometry scientifically. All known space-relations are measurable, that is, they may be brought to determination of magnitudes (lines, angles, surfaces, volumes). Problems in geometry can therefore be solved, by finding methods of calculation for arriving at unknown magnitudes from known ones. This is done in *analytical geometry*, where all forms of space are treated only as quantities and determined by means of other quantities. Even the axioms themselves make reference to magnitudes. The straight line is defined as the *shortest* between two points, which is a determination of quantity. The axiom of parallels declares that if two straight lines in a plane do not intersect (are parallel), the alternate angles, or the corresponding angles, made by a third line intersecting them, are equal; or it may be laid down instead that the sum of the angles of any triangle is equal to two right angles. These, also, are determinations of quantity.

Now we may start with this view of space, according to which the position of a point may be determined by measurements in relation to any given figure (system of co-ordinates), taken as fixed, and then inquire what are the special characteristics of our space as manifested in the measurements that have to be made, and how it differs from other extended quantities of like variety. This path was first entered by one too early lost to science, B. Riemann of Göttingen.[1] It has the peculiar advantage that all its operations consist in pure calculation of quantities, which quite obviates the danger of habitual perceptions being taken for necessities of thought.

The number of measurements necessary to give the position of a point, is equal to the number of dimensions of the space in question. In a line the distance from one fixed point is sufficient, that is to say, one quantity; in a surface the distances from two fixed points must be given; in space, the distances from three; or we require, as on the earth, longitude, latitude, and height above the sea, or, as is usual in analytical geometry, the distances from three co-ordinate planes. Riemann

[1] Ueber die Hypothesen welche der Geometrie zu Grunde liegen, Habilitationsschrift vom 10 Juni 1854. (*Abhandl. der königl. Gesellsch. zu Göttingen*, Bd. XIII.)

calls a system of differences in which one thing can be determined by n measurements an '$n$fold extended aggregate' or an 'aggregate of n dimensions.' Thus the space in which we live is a threefold, a surface is a twofold, and a line is a simple extended aggregate of points. Time also is an aggregate of one dimension. The system of colours is an aggregate of three dimensions, inasmuch as each colour, according to the investigations of Thomas Young and of Clerk Maxwell,[1] may be represented as a mixture of three primary colours, taken in definite quantities. The particular mixtures can be actually made with the colour-top.

In the same way we may consider the system of simple tones[2] as an aggregate of two dimensions, if we distinguish only pitch and intensity, and leave out of account differences of timbre. This generalisation of the idea is well suited to bring out the distinction between space of three dimensions and other aggregates. We can, as we know from daily experience, compare the vertical distance of two points with the horizontal distance of two others, because we can apply a measure first to the one pair and then to the other. But we cannot compare the difference between two tones of equal pitch and different intensity, with that between two tones of equal intensity and different pitch. Riemann showed, by considerations of this kind, that the essential foun-

---

[1] Helmholtz's *Popular Lectures*, Series I. p. 243.   [2] Ibid. p. 86.

dation of any system of geometry, is the expression that it gives for the distance between two points lying in any direction towards one another, beginning with the infinitesimal interval. He took from analytical geometry the most general form for this expression, that, namely, which leaves altogether open the kind of measurements by which the position of any point is given.[1] Then he showed that the kind of free mobility without change of form which belongs to bodies in our space can only exist when certain quantities yielded by the calculation [2] — quantities that coincide with Gauss's measure of surface-curvature when they are expressed for surfaces — have everywhere an equal value. For this reason Riemann calls these quantities, when they have the same value in all directions for a particular spot, the measure of curvature of the space at this spot. To prevent misunderstanding,[3] I will once more observe that this so-called measure of space-curvature is a quantity obtained by purely analytical calculation, and that its introduction involves no suggestion of relations that would have a meaning only for sense-perception. The name is merely taken,

[1] For the square of the distance of two infinitely near points the expression is a homogeneous quadric function of the differentials of their co-ordinates.

[2] They are algebraical expressions compounded from the coefficients of the various terms in the expression for the square of the distance of two contiguous points and from their differential quotients.

[3] As occurs, for instance, in the above-mentioned work of Tobias, pp. 70, etc.

as a short expression for a complex relation, from the one case in which the quantity designated admits of sensible representation.

Now whenever the value of this measure of curvature in any space is everywhere zero, that space everywhere conforms to the axioms of Euclid; and it may be called a *flat* (*homaloid*) space in contradistinction to other spaces, analytically constructible, that may be called *curved*, because their measure of curvature has a value other than zero. Analytical geometry may be as completely and consistently worked out for such spaces as ordinary geometry can for our actually existing homaloid space.

If the measure of curvature is positive we have *spherical* space, in which straightest lines return upon themselves and there are no parallels. Such a space would, like the surface of a sphere, be unlimited but not infinitely great. A constant negative measure of curvature on the other hand gives *pseudo-spherical* space, in which straightest lines run out to infinity, and a pencil of straightest lines may be drawn, in any flattest surface, through any point which does not intersect another given straightest line in that surface.

Beltrami [1] has rendered these last relations imaginable by showing that the points, lines, and surfaces of a pseudospherical space of three dimensions, can be so

[1] *Teoria fondamentale, &c., ut sup.*

portrayed in the interior of a sphere in Euclid's homaloid space, that every straightest line or flattest surface of the pseudospherical space is represented by a straight line or a plane, respectively, in the sphere. The surface itself of the sphere corresponds to the infinitely distant points of the pseudospherical space; and the different parts of this space, as represented in the sphere, become smaller, the nearer they lie to the spherical surface, diminishing more rapidly in the direction of the radii than in that perpendicular to them. Straight lines in the sphere, which only intersect beyond its surface, correspond to straightest lines of the pseudospherical space which never intersect.

Thus it appeared that space, considered as a region of measurable quantities, does not at all correspond with the most general conception of an aggregate of three dimensions, but involves also special conditions, depending on the perfectly free mobility of solid bodies without change of form to all parts of it and with all possible changes of direction; and, further, on the special value of the measure of curvature which for our actual space equals, or at least is not distinguishable from, zero. This latter definition is given in the axioms of straight lines and parallels.

Whilst Riemann entered upon this new field from the side of the most general and fundamental questions of analytical geometry, I myself arrived at similar conclusions,[1] partly from seeking to represent in space the system of colours, involving the comparison of one threefold extended aggregate with another, and partly from inquiries on the origin of our ocular measure for distances in the field of vision. Riemann starts by assuming the above-mentioned algebraical expression which represents in the most general form the distance between two infinitely near points, and deduces therefrom, the conditions of mobility of rigid figures. I, on the other hand, starting from the observed fact that the movement of rigid figures is possible in our space, with the degree of freedom that we know, deduce the necessity of the algebraic expression taken by Riemann as an axiom. The assumptions that I had to make as the basis of the calculation were the following.

First, to make algebraical treatment at all possible, it must be assumed that the position of any point A can be determined, in relation to certain given figures taken as fixed bases, by measurement of some kind of magnitudes, as lines, angles between lines, angles between surfaces, and so forth. The measurements necessary for determining the position of A are known as its co-ordinates. In general, the number of co-ordinates necessary for the complete determination of the position of a point, marks the number of the dimen-

[1] Ueber die Thatsachen die der Geometrie zum Grunde liegen (*Nachrichten von der königl. Ges. d. Wiss. zu Göttingen*, Juni 3, 1868).

sions of the space in question. It is further assumed that with the movement of the point A, the magnitudes used as co-ordinates vary continuously.

Secondly, the definition of a solid body, or rigid system of points, must be made in such a way as to admit of magnitudes being compared by congruence. As we must not, at this stage, assume any special methods for the measurement of magnitudes, our definition can, in the first instance, run only as follows: Between the co-ordinates of any two points belonging to a solid body, there must be an equation which, however the body is moved, expresses a constant spatial relation (proving at last to be the distance) between the two points, and which is the same for congruent pairs of points, that is to say, such pairs as can be made successively to coincide in space with the same fixed pair of points.

However indeterminate in appearance, this definition involves most important consequences, because with increase in the number of points, the number of equations increases much more quickly than the number of co-ordinates which they determine. Five points, A, B, C, D, E, give ten different pairs of points

AB, AC, AD, AE,
BC, BD, BE,
CD, CE,
DE,

and therefore ten equations, involving in space of three dimensions fifteen variable co-ordinates. But of these fifteen, six must remain arbitrary, if the system of five points is to admit of free movement and rotation, and thus the ten equations can determine only nine co-ordinates as functions of the six variables. With six points we obtain fifteen equations for twelve quantities, with seven points twenty-one equations for fifteen, and so on. Now from n independent equations we can determine n contained quantities, and if we have more than n equations, the superfluous ones must be deducible from the first n. Hence it follows that the equations which subsist between the co-ordinates of each pair of points of a solid body must have a special character, seeing that, when in space of three dimensions they are satisfied for nine pairs of points as formed out of any five points, the equation for the tenth pair follows by logical consequence. Thus our assumption for the definition of solidity, becomes quite sufficient to determine the kind of equations holding between the co-ordinates of two points rigidly connected.

Thirdly, the calculation must further be based on the fact of a peculiar circumstance in the movement of solid bodies, a fact so familiar to us that but for this inquiry it might never have been thought of as something that need not be. When in our space of three dimensions two points of a solid body are kept fixed, its movements are limited to rotations round the straight line connecting them. If we turn it completely round once, it again occupies exactly the position it had at first. This fact, that rotation in one direction always brings a solid body back into its original position, needs special mention. A system of geometry is possible without it. This is most easily seen in the geometry of a plane. Suppose that with every rotation of a plane figure its linear dimensions increased in proportion to the angle of rotation, the figure after one whole rotation through 360 degrees would no longer coincide with itself as it was originally. But any second figure that was congruent with the first in its original position might be made to coincide with it in its second position by being also turned through 360 degrees. A consistent system of geometry would be possible upon this supposition, which does not come under Riemann's formula.

On the other hand I have shown that the three assumptions taken together form a sufficient basis for the starting-point of Riemann's investigation, and thence for all his further results relating to the distinction of different spaces according to their measure of curvature.

It still remained to be seen whether the laws of motion, as dependent on moving forces, could also be consistently transferred to spherical or pseudospherical space. This investigation has been carried out by Professor Lipschitz of Bonn.[1] It is found that the comprehensive expression for all the laws of dynamics, Hamilton's principle, may be directly transferred to spaces of which the measure of curvature is other than zero. Accordingly, in this respect also, the disparate systems of geometry lead to no contradiction.

We have now to seek an explanation of the special characteristics of our own flat space, since it appears that they are not implied in the general notion of an extended quantity of three dimensions and of the free mobility of bounded figures therein. *Necessities of thought*, such as are involved in the conception of such a variety, and its measurability, or from the most general of all ideas of a solid figure contained in it, and of its free mobility, they undoubtedly are not. Let us then examine the opposite assumption as to their origin being empirical, and see if they can be inferred from facts of experience and so established, or if, when tested by experience, they are perhaps to be rejected. If they are of empirical origin, we must be able to represent to ourselves connected series of facts, indicating a different value for the measure of curvature from that of Euclid's flat space. But if we can

---

[1] 'Untersuchungen über die ganzen homogenen Functionen von n Differentialen' (Borchardt's *Journal für Mathematik*, Bd. lxx. 3, 71; lxxiii. 3, 1); ' Untersuchung eines Problems der Variationsrechnung' (*Ibid.* Bd. lxxiv.).

imagine such spaces of other sorts, it cannot be maintained that the axioms of geometry are necessary consequences of an *à priori* transcendental form of intuition, as Kant thought.

The distinction between spherical, pseudospherical, and Euclid's geometry depends, as was above observed, on the value of a certain constant called, by Riemann, the measure of curvature of the space in question. The value must be zero for Euclid's axioms to hold good. If it were not zero, the sum of the angles of a large triangle would differ from that of the angles of a small one, being larger in spherical, smaller in pseudospherical, space. Again, the geometrical similarity of large and small solids or figures is possible only in Euclid's space. All systems of practical mensuration that have been used for the angles of large rectilinear triangles, and especially all systems of astronomical measurement which make the parallax of the immeasurably distant fixed stars equal to zero (in pseudospherical space the parallax even of infinitely distant points would be positive), confirm empirically the axiom of parallels, and show the measure of curvature of our space thus far to be indistinguishable from zero. It remains, however, a question, as Riemann observed, whether the result might not be different if we could use other than our limited base-lines, the greatest of which is the major axis of the earth's orbit.

Meanwhile, we must not forget that all geometrical measurements rest ultimately upon the principle of congruence. We measure the distance between points by applying to them the compass, rule, or chain. We measure angles by bringing the divided circle or theodolite to the vertex of the angle. We also determine straight lines by the path of rays of light which in our experience is rectilinear; but that light travels in shortest lines as long as it continues in a medium of constant refraction would be equally true in space of a different measure of curvature. Thus all our geometrical measurements depend on our instruments being really, as we consider them, invariable in form, or at least on their undergoing no other than the small changes we know of, as arising from variation of temperature, or from gravity acting differently at different places.

In measuring, we only employ the best and surest means we know of to determine, what we otherwise are in the habit of making out by sight and touch or by pacing. Here our own body with its organs is the instrument we carry about in space. Now it is the hand, now the leg, that serves for a compass, or the eye turning in all directions is our theodolite for measuring arcs and angles in the visual field.

Every comparative estimate of magnitudes or measurement of their spatial relations proceeds therefore upon a supposition as to the behaviour of certain physical things, either the human body or other instruments employed. The supposition may be in the highest degree probable and in closest harmony with all other physical relations known to us, but yet it passes beyond the scope of pure space-intuition.

It is in fact possible to imagine conditions for bodies apparently solid such that the measurements in Euclid's space become what they would be in spherical or pseudospherical space. Let me first remind the reader that if all the linear dimensions of other bodies, and our own, at the same time were diminished or increased in like proportion, as for instance to half or double their size, we should with our means of space-perception be utterly unaware of the change. This would also be the case if the distension or contraction were different in different directions, provided that our own body changed in the same manner, and further that a body in rotating assumed at every moment, without suffering or exerting mechanical resistance, the amount of dilatation in its different dimensions corresponding to its position at the time. Think of the image of the world in a convex mirror. The common silvered globes set up in gardens give the essential features, only distorted by some optical irregularities. A well-made convex mirror of moderate aperture represents the objects in front of it as apparently solid and in fixed positions behind its surface. But the images of the distant horizon and of the sun in the sky lie behind the mirror at a limited distance, equal to its focal length. Between these and the surface of the mirror are found the images of all the other objects before it, but the images are diminished and flattened in proportion to the distance of their objects from the mirror. The flattening, or decrease in the third dimension, is relatively greater than the decrease of the surface-dimensions. Yet every straight line or every plane in the outer world is represented by a straight line or a plane in the image. The image of a man measuring with a rule a straight line from the mirror would contract more and more the farther he went, but with his shrunken rule the man in the image would count out exactly the same number of centimetres as the real man. And, in general, all geometrical measurements of lines or angles made with regularly varying images of real instruments would yield exactly the same results as in the outer world, all congruent bodies would coincide on being applied to one another in the mirror as in the outer world, all lines of sight in the outer world would be represented by straight lines of sight in the mirror. In short I do not see how men in the mirror are to discover that their bodies are not rigid solids and their experiences good examples of the correctness of Euclid's axioms. But if they could look out upon our world as we can look into theirs, without overstepping the boundary, they must declare it to be a picture in a spherical mirror, and would speak of us just as we speak of them; and if two inhabitants of the different worlds could communicate with one another, neither, so far as I can see, would be able to convince the other that he had the true, the other the distorted, relations. Indeed I cannot see that such a question would have any meaning at all, so long as mechanical considerations are not mixed up with it.

Now Beltrami's representation of pseudospherical space in a sphere of Euclid's space, is quite similar, except that the background is not a plane as in the convex mirror, but the surface of a sphere, and that the proportion in which the images as they approach the spherical surface contract, has a different mathematical expression.[1] If we imagine then, conversely, that in the sphere, for the interior of which Euclid's axioms hold good, moving bodies contract as they depart from the centre like the images in a convex mirror, and in such a way that their representatives in pseudospherical space retain their dimensions unchanged,—observers whose bodies were regularly subjected to the same change would obtain the same results from the geometrical measurements they could make as if they lived in pseudospherical space.

[1] Compare the Appendix at the end of this Lecture.

We can even go a step further, and infer how the objects in a pseudospherical world, were it possible to enter one, would appear to an observer, whose eye-measure and experiences of space had been gained like ours in Euclid's space. Such an observer would continue to look upon rays of light or the lines of vision as straight lines, such as are met with in flat space, and as they really are in the spherical representation of pseudospherical space. The visual image of the objects in pseudospherical space would thus make the same impression upon him as if he were at the centre of Beltrami's sphere. He would think he saw the most remote objects round about him at a finite distance,[1] let us suppose a hundred feet off. But as he approached these distant objects, they would dilate before him, though more in the third dimension than superficially, while behind him they would contract. He would know that his eye judged wrongly. If he saw two straight lines which in his estimate ran parallel for the hundred feet to his world's end, he would find on following them that the farther he advanced the more they diverged, because of the dilatation of all the objects to which he approached. On the other hand, behind him, their distance would seem to diminish, so that as he advanced they would

[1] The reciprocal of the square of this distance, expressed in negative quantity, would be the measure of curvature of the pseudospherical space.

appear always to diverge more and more. But two straight lines which from his first position seemed to converge to one and the same point of the background a hundred feet distant, would continue to do this however far he went, and he would never reach their point of intersection.

Now we can obtain exactly similar images of our real world, if we look through a large convex lens of corresponding negative focal length, or even through a pair of convex spectacles if ground somewhat prismatically to resemble pieces of one continuous larger lens. With these, like the convex mirror, we see remote objects as if near to us, the most remote appearing no farther distant than the focus of the lens. In going about with this lens before the eyes, we find that the objects we approach dilate exactly in the manner I have described for pseudospherical space. Now any one using a lens, were it even so strong as to have a focal length of only sixty inches, to say nothing of a hundred feet, would perhaps observe for the first moment that he saw objects brought nearer. But after going about a little the illusion would vanish, and in spite of the false images he would judge of the distances rightly. We have every reason to suppose that what happens in a few hours to any one beginning to wear spectacles would soon enough be experienced in pseudospherical space. In short, pseudospherical space would not seem to us very strange, comparatively speaking; we should only at first be subject to illusions in measuring by eye the size and distance of the more remote objects.

There would be illusions of an opposite description, if, with eyes practised to measure in Euclid's space, we entered a spherical space of three dimensions. We should suppose the more distant objects to be more remote and larger than they are, and should find on approaching them that we reached them more quickly than we expected from their appearance. But we should also see before us objects that we can fixate only with diverging lines of sight, namely, all those at a greater distance from us than the quadrant of a great circle. Such an aspect of things would hardly strike us as very extraordinary, for we can have it even as things are if we place before the eye a slightly prismatic glass with the thicker side towards the nose: the eyes must then become divergent to take in distant objects. This excites a certain feeling of unwonted strain in the eyes, but does not perceptibly change the appearance of the objects thus seen. The strangest sight, however, in the spherical world would be the back of our own head, in which all visual lines not stopped by other objects would meet again, and which must fill the extreme background of the whole perspective picture.

At the same time it must be noted that as a small elastic flat disk, say of india-rubber, can only be fitted to a slightly curved spherical surface with relative contraction of its border and distension of its centre, so our bodies, developed in Euclid's flat space, could not pass into curved space without undergoing similar distensions and contractions of their parts, their coherence being of course maintained only in as far as their elasticity permitted their bending without breaking. The kind of distension must be the same as in passing from a small body imagined at the centre of Beltrami's sphere to its pseudospherical or spherical representation. For such passage to appear possible, it will always have to be assumed that the body is sufficiently elastic and small in comparison with the real or imaginary radius of curvature of the curved space into which it is to pass.

These remarks will suffice to show the way in which we can infer from the known laws of our sensible perceptions the series of sensible impressions which a spherical or pseudospherical world would give us, if it existed. In doing so, we nowhere meet with inconsistency or impossibility any more than in the calculation of its metrical proportions. We can represent to ourselves the look of a pseudospherical world in all directions just as we can develop the conception of it. Therefore it cannot be allowed that the axioms of our geometry depend on the native form of our perceptive faculty, or are in any way connected with it.

It is different with the three dimensions of space. As all our means of sense-perception extend only to space of three dimensions, and a fourth is not merely a modification of what we have, but something perfectly new, we find ourselves by reason of our bodily organisation quite unable to represent a fourth dimension.

In conclusion, I would again urge that the axioms of geometry are not propositions pertaining only to the pure doctrine of space. As I said before, they are concerned with quantity. We can speak of quantities only when we know of some way by which we can compare, divide, and measure them. All space-measurements, and therefore in general all ideas of quantities applied to space, assume the possibility of figures moving without change of form or size. It is true we are accustomed in geometry to call such figures purely geometrical solids, surfaces, angles, and lines, because we abstract from all the other distinctions, physical and chemical, of natural bodies; but yet one physical quality, rigidity, is retained. Now we have no other mark of rigidity of bodies or figures but congruence, whenever they are applied to one another at any time or place, and after any revolution. We cannot, however, decide by pure geometry, and without mechanical considerations, whether the coinciding bodies may not both have varied in the same sense.

If it were useful for any purpose, we might with perfect consistency look upon the space in which we live as the apparent space behind a convex mirror with its shortened and contracted background; or we might consider a bounded sphere of our space, beyond the limits of which we perceive nothing further, as infinite pseudospherical space. Only then we should have to ascribe to the bodies which appear to us to be solid, and to our own body at the same time, corresponding distensions and contractions, and we should have to change our system of mechanical principles entirely; for even the proposition that every point in motion, if acted upon by no force, continues to move with unchanged velocity in a straight line, is not adapted to the image of the world in the convex-mirror. The path would indeed be straight, but the velocity would depend upon the place.

Thus the axioms of geometry are not concerned with space-relations only but also at the same time with the mechanical deportment of solidest bodies in motion. The notion of rigid geometrical figure might indeed be conceived as transcendental in Kant's sense, namely, as formed independently of actual experience, which need not exactly correspond therewith, any more than natural bodies do ever in fact correspond exactly to the abstract notion we have obtained of them by induction. Taking the notion of rigidity thus as a mere ideal, a strict Kantian might certainly look upon the geometrical axioms as propositions given, à *priori*, by transcendental intuition, which no experience could either confirm or refute, because it must first be decided by them whether any natural bodies can be considered as rigid. But then we should have to maintain that the axioms of geometry are not synthetic propositions, as Kant held them; they would merely define what qualities and deportment a body must have to be recognised as rigid.

But if to the geometrical axioms we add propositions relating to the mechanical properties of natural bodies, were it only the axiom of inertia, or the single proposition, that the mechanical and physical properties of bodies and their mutual reactions are, other circumstances remaining the same, independent of place, such a system of propositions has a real import which can be confirmed or refuted by experience, but just for the same reason can also be gained by experience. The mechanical axiom, just cited, is in fact of the utmost importance for the whole system of our mechanical and physical conceptions. That rigid solids, as we call them, which are really nothing else than elastic solids of great resistance, retain the same form in every part of space if no external force affects them, is a single case falling under the general principle.

In conclusion, I do not, of course, maintain that mankind first arrived at space-intuitions, in agreement with the axioms of Euclid, by any carefully executed systems of exact measurement. It was rather a succession of everyday experiences, especially the perception of the geometrical similarity of great and small bodies, only possible in flat space, that led to the rejection, as impossible, of every geometrical representation at variance with this fact. For this no knowledge of the necessary logical connection between the observed fact of geometrical similarity and the axioms was needed; but only an intuitive apprehension of the typical relations between lines, planes, angles, &c., obtained by numerous and attentive observations—an intuition of the kind the artist possesses of the objects he is to represent, and by means of which he decides with certainty and accuracy whether a new combination, which he tries, will correspond or not with their nature. It is true that we have no word but *intuition* to mark this; but it is knowledge empirically gained by the aggregation and reinforcement of similar recurrent impressions in memory, and not a transcendental form given before experience. That other such empirical intuitions of fixed typical relations, when not clearly comprehended, have frequently enough been taken by metaphysicians for *à priori* principles, is a point on which I need not insist.

To sum up, the final outcome of the whole inquiry may be thus expressed:—

(1.) The axioms of geometry, taken by themselves out of all connection with mechanical propositions, represent no relations of real things. When thus isolated, if we regard them with Kant as forms of intuition transcendentally given, they constitute a form into which any empirical content whatever will fit, and which therefore does not in any way limit or determine beforehand the nature of the content. This is true, however, not only of Euclid's axioms, but also of the axioms of spherical and pseudospherical geometry.

(2.) As soon as certain principles of mechanics are conjoined with the axioms of geometry, we obtain a system of propositions which has real import, and which can be verified or overturned by empirical observations, just as it can be inferred from experience. If such a system were to be taken as a transcendental form of intuition and thought, there must be assumed a pre-established harmony between form and reality.

APPENDIX.

The elements of the geometry of spherical space are most easily obtained by putting for space of four dimensions the equation for the sphere

$$x^2 + y^2 + z^2 + t^2 = R^2 \quad \ldots \ldots \ldots (1.)$$

and for the distance ds between the points (x, y, z, t) and $[(x+dx)\,(y+dy)\,(z+dz)\,(t+dt)]$ the value

$$ds^2 = dx^2 + dy^2 + dz^2 + dt^2 \quad \ldots \ldots (2.)$$

It is easily found by means of the methods used for three dimensions that the shortest lines are given by equations of the form

$$\left. \begin{array}{l} ax + by + cz + ft = 0 \\ ax + \beta y + \gamma z + \phi t = 0 \end{array} \right\} \quad \ldots \ldots \ldots (3.)$$

in which a, b, c, f, as well as a, β, γ, ϕ, are constants.

The length of the shortest arc, s, between the points (x, y, z, t), and (ξ, η, ζ, τ) follows, as in the sphere, from the equation

$$\cos \frac{s}{R} = \frac{x\xi + y\eta + z\zeta + t\tau}{R^2} \quad \ldots \ldots \ldots (4.)$$

One of the co-ordinates may be eliminated from the values given in 2 to 4, by means of equation 1, and the expressions then apply to space of three dimensions.

If we take the distances from the points

$$\xi = \eta = \zeta = 0$$

from which equation 1 gives $\tau = R$, then,

$$\sin\left(\frac{s_0}{R}\right) = \frac{\sigma}{R}$$

in which $\quad \sigma = \sqrt{x^2 + y^2 + z^2}$

or, $\quad s_0 = R \cdot \arcsin\left(\frac{\sigma}{R}\right) = R \cdot \operatorname{arc\,tang}\left(\frac{\sigma}{t}\right) \ldots$ (5.)

In this, s_0 is the distance of the point x, y, z, measured from the centre of the co-ordinates.

If now we suppose the point x, y, z, of spherical space, to be projected in a point of plane space whose co-ordinates are respectively

$$\mathfrak{x} = \frac{Rx}{t} \quad \mathfrak{y} = \frac{Ry}{t} \quad \mathfrak{z} = \frac{Rz}{t}$$

$$\mathfrak{x}^2 + \mathfrak{y}^2 + \mathfrak{z}^2 = \mathfrak{r}^2 = \frac{R^2 \sigma^2}{t^2}$$

then in the plane space the equations 3, which belong to the straightest lines of spherical space, are equations of the straight line. Hence the shortest lines of spherical space are represented in the system of $\mathfrak{x}$, $\mathfrak{y}$, $\mathfrak{z}$, by straight lines. For very small values of x, y, z, $t = R$, and

$$\mathfrak{x} = x, \quad \mathfrak{y} = y, \quad \mathfrak{z} = z$$

Immediately about the centre of the co-ordinates, the measurements of both spaces coincide. On the other hand, we have for the distances from the centre

$$s_0 = R \cdot \operatorname{arc\,tang}\left(\pm \frac{\mathfrak{r}}{R}\right) \ldots \text{(6.)}$$

In this, $\mathfrak{r}$ may be infinite; but every point of plane space must be the projection of two points of the sphere, one for which $s_0 < \frac{1}{2} R\pi$, and one for which $s_0 > \frac{1}{2} R\pi$. The extension in the direction of $\mathfrak{r}$ is then

$$\frac{ds_0}{d\mathfrak{r}} = \frac{R^2}{R^2 + \mathfrak{r}^2}$$

APPENDIX.

In order to obtain corresponding expressions for pseudo-spherical space, let R and t be imaginary; that is, $R = \Re i$, and $t = t i$. Equation 6 gives then

$$\tan \frac{s_0}{\Re i} = \pm \frac{\mathfrak{r}}{\Re i}$$

from which, eliminating the imaginary form, we get

$$s_0 = \tfrac{1}{2}\, \Re \; \text{log. nat.}\; \frac{\Re + \mathfrak{r}}{\Re - \mathfrak{r}}$$

Here s_0 has real values only as long as $\mathfrak{r} = R$; for $\mathfrak{r} = \Re$ the distance s_0 in pseudospherical space is infinite. The image in plane space is, on the contrary, contained in the sphere of radius R, and every point of this sphere forms only one point of the infinite pseudospherical space. The extension in the direction of $\mathfrak{r}$ is

$$\frac{ds_0}{d\mathfrak{r}} = \frac{\Re^2}{\Re^2 - \mathfrak{r}^2}$$

For linear elements, on the contrary, whose direction is at right angles to $\mathfrak{r}$, and for which t is unchanged, we have in both cases

$$\frac{\sqrt{dx^2 + dy^2 + dz^2}}{\sqrt{d\mathfrak{x}^2 + d\mathfrak{y}^2 + d\mathfrak{z}^2}} = \frac{t}{R} = \frac{t}{\Re} = \frac{\sigma}{\mathfrak{r}}$$
$$= \frac{\sqrt{x^2 + y^2 + z^2}}{\sqrt{\mathfrak{x}^2 + \mathfrak{y}^2 + \mathfrak{z}^2}}$$

ON

THE RELATION OF OPTICS

TO

PAINTING.

*Being the substance of a series of Lectures delivered in
Cologne, Berlin, and Bonn.*

I FEAR that the announcement of my intention to address you on the subject of plastic art may have created no little surprise among some of my hearers. For I cannot doubt that many of you have had more frequent opportunities of viewing works of art, and have more thoroughly studied its historical aspects, than I can lay claim to have done; or indeed have had personal experience in the actual practice of art, in which I am entirely wanting. I have arrived at my artistic studies by a path which is but little trod, that is, by the physiology of the senses; and in reference to those who have a long acquaintance with, and who are quite at home in the beautiful fields of art, I may compare myself to a traveller who has entered upon them by a steep and stony mountain path, but who, in doing so, has passed many a stage from which a good point of view is obtained. If therefore I relate to you what I consider I have observed, it is with the understanding that I wish to regard myself as open to instruction by those more experienced than myself.

The physiological study of the manner in which the perceptions of our senses originate, how impressions from without pass into our nerves, and how the condition of the latter is thereby altered, presents many points of contact with the theory of the fine arts. On a former occasion I endeavoured to establish such a relation between the physiology of the sense of hearing, and the theory of music. Those relations in that case are particularly clear and distinct, because the elementary forms of music depend more closely on the nature and on the peculiarities of our perceptions than is the case in other arts, in which the nature of the material to be used and of the objects to be represented has a far greater influence. Yet even in those other branches of art, the especial mode of perception of that organ of sense by which the impression is taken up is not without importance; and a theoretical insight into its action, and into the principle of its methods, cannot be complete if this physiological element is not taken into account. Next to music this seems to predominate more particularly in painting, and this is the reason why I have chosen painting as the subject of my present lecture.

The more immediate object of the painter is to produce in us by his palette a lively visual impression of the objects which he has endeavoured to represent. The aim, in a certain sense, is to produce a kind of optical illusion; not indeed that, like the birds who pecked at the painted grapes of Apelles, we are to suppose we have present the real objects themselves, and not a picture; but in so far that the artistic representation produces in us a conception of their objects as vivid and as powerful as if we had them actually before us. The study of what are called illusions of the senses is however a very prominent and important part of the physiology of the senses; for just those cases in which external impressions evoke conceptions which are not in accordance with reality are particularly instructive for discovering the laws of those means and processes by which normal perceptions originate. We must look upon artists as persons whose observation of sensuous impressions is particularly vivid and accurate, and whose memory for these images is particularly true. That which long tradition has handed down to the men most gifted in this respect, and that which they have found by innumerable experiments in the most varied directions, as regards means and methods of representation, forms a series of important and significant facts, which the physiologist, who has here to learn from the artist, cannot afford to neglect. The study of works of art will throw great light on the question as to which elements and relations of our visual impressions are most predominant in determining our conception of what is seen, and what others are of less importance. As far as lies within his power, the artist will seek to foster the former at the cost of the latter.

In this sense then a careful observation of the works of the great masters will be serviceable, not only to physiological optics, but also because the investigation of the laws of the perceptions and of the observations of the senses will promote the theory of art, that is, the comprehension of its mode of action.

We have not here to do with a discussion of the ultimate objects and aims of art, but only with an examination of the action of the elementary means with which it works. The knowledge of the latter must, however, form an indispensable basis for the solution of the deeper questions, if we are to understand the problems which the artist has to solve, and the mode in which he attempts to attain his object.

I need scarcely lay stress on the fact, following as it does from what I have already said, that it is not my intention to furnish instructions according to which the artist is to work. I consider it a mistake to suppose that any kind of æsthetic lectures such as these can ever do so; but it is a mistake which those very frequently make who have only practical objects in view.

I. Form.

The painter seeks to produce in his picture an image of external objects. The first aim of our investigation must be to ascertain what degree and what kind of similarity he can expect to attain, and what limits are assigned to him by the nature of his method. The uneducated observer usually requires nothing more than an illusive resemblance to nature: the more this is obtained, the more does he delight in the picture. An observer, on the contrary, whose taste in works of art has been more finely educated, will, consciously or unconsciously, require something more, and something different. A faithful copy of crude Nature he will at most regard as an artistic feat. To satisfy him, he will need artistic selection, grouping, and even idealisation of the objects represented. The human figures in a work of art must not be the everyday figures, such as we see in photographs; they must have expression, and a characteristic development, and if possible beautiful forms, which have perhaps belonged to no living individuals or indeed any individuals which ever have existed, but only to such a one as might exist, and as must exist, to produce a vivid perception of any particular aspect of human existence in its complete and unhindered development.

If however the artist is to produce an artistic arrangement of only idealised types, whether of man or of natural objects, must not the picture be an actual, complete, and directly true delineation of that which would appear if it anywhere came into being?

Since the picture is on a plane surface, this faithful representation can of course only give a faithful perspective view of the objects. Yet our eye, which in its optical properties is equivalent to a camera obscura, the well-known apparatus of the photographer, gives on the retina, which is its sensitive plate, only perspective views of the external world; these are stationary, like the drawing on a picture, as long as the standpoint of the eye is not altered. And, in fact, if we restrict ourselves in the first place to the form of the object viewed, and disregard for the present any consideration of colour, by a correct perspective drawing we can present to the eye of an observer, who views it from a correctly chosen point of view, the same forms of the visual image as the inspection of the objects themselves would present to the same eye, when viewed from the corresponding point of view.

But apart from the fact that any movement of the observer, whereby his eye changes its position, will produce displacements of the visual image, different when he stands before objects from those when he stands before the image, I could speak of only *one* eye for which equality of impression is to be established. We however see the world with *two* eyes, which occupy somewhat different positions in space, and which therefore show two different perspective views of objects before us. This difference of the images of the two eyes forms one of the most important means of estimating the distance of objects from our eye, and of estimating depth, and this is what is wanting to the painter, or even turns against him; since in binocular vision the picture distinctly forces itself on our perception as a plane surface.

You must all have observed the wonderful vividness which the solid form of objects acquires when good stereoscopic images are viewed in the stereoscope, a kind of vividness in which either of the pictures is wanting when viewed without the stereoscope. The illusion is most striking and instructive with figures in simple line; models of crystals and the like, in which there is no other element of illusion. The reason of this deception is, that looking with two eyes we view the world simultaneously from somewhat different points of view, and thereby acquire two different perspective images. With the right eye we see somewhat more of the right side of objects before us, and also somewhat more of those behind it, than we do with the left eye; and conversely we see with the left, more of the left side of an object, and of the background behind its left edges, and partially concealed by the edge. But a flat picture shows to the right eye absolutely the same picture, and all objects represented upon it, as to the left eye. If then we make for each eye such a picture as that eye would perceive if itself looked at the object, and if both pictures are combined in the stereoscope, so that each eye sees its corresponding picture, then as far as form is concerned the same impression is produced in the two eyes as the object itself produces. But if we look at a drawing or a picture with both eyes, we just as easily recognise that it is a representation on a plane surface, which is different from that which the actual object would show simultaneously to both eyes. Hence is due the well-known increase in the vividness of a picture if it is looked at with only one eye, and while quite stationary, through a dark tube; we thus exclude any comparison of its distance with that of adjacent objects in the room. For it must be observed that as we use different pictures seen with the two eyes for the perception of depth, in like manner as the body moves from one place to another, the pictures seen by the same eye serve for the same purpose. In moving, whether on foot or riding, the nearer objects are apparently displaced in comparison with the more distant ones; the former appear to recede, the latter appear to move with us. Hence arises a far stricter distinction between what is near and what is distant, than seeing with one eye from one and the same spot would ever afford us. If we move towards the picture, the sensuous impression that it is a flat picture hanging against the wall forces itself more strongly upon us than if we look at it while we are stationary. Compared with a large picture at a greater distance, all those elements which depend on binocular vision and on the movement of the body are less operative, because in very distant objects the differences between the images of the two eyes, or between the aspect from adjacent points of view, seem less. Hence large pictures furnish a less distorted aspect of their object than small ones, while the impression on a stationary eye, of a small picture close at hand, might be just the same as that of a large distant one. In a painting close at hand, the fact that it is a flat picture continually forces itself more powerfully and more distinctly on our perception.

The fact that perspective drawings, which are taken from too near a point of view, may easily produce a distorted impression, is, I think, connected with this. For here the want of the second representation for the other eye, which would be very different, is too marked. On the other hand, what are called geometrical projections, that is, perspective drawings which represent a view taken from an infinite distance, give in many cases a particularly favourable view of the object, although they correspond to a point of sight which does not in reality occur. Here the pictures of both eyes for such an object are the same.

You will notice that in these respects there is a primary incongruity, and one which cannot be got over, between the aspect of a picture and the aspect of reality. This incongruity may be lessened, but never entirely overcome. Owing to the imperfect action of binocular vision, the most important natural means is lost of enabling the observer to estimate the depth of objects represented in the picture. The painter possesses a series of subordinate means, partly of limited applicability, and partly of slight effect, of expressing various distances by depth. It is not unimportant to become acquainted with these elements, as arising out of theoretical considerations; for in the practice of the art of painting they have manifestly exercised great influence on the arrangement, selection, and mode of illumination of the objects represented. The distinctness of what is represented is indeed of subordinate importance when considered in reference to the ideal aims of art; it must not however be depreciated, for it is the first condition by which the observer attains an intelligibility of expression, which impresses itself without fatigue on the observer.

This direct intelligibility is again the preliminary condition for an undisturbed, and vivid action of the picture on the feelings and mood of the observer.

The subordinate methods of expressing depth which have been referred to, depend in the first place on perspective. Nearer objects partially conceal more distant ones, but can never themselves be concealed by the latter. If therefore the painter skilfully groups his objects, so that the feature in question comes into play, this gives at once a very certain gradation of far and near. This mutual concealment may even preponderate over the binocular perception of depth, if stereoscopic pictures are intentionally produced in which each counteracts the other. Moreover, in bodies of regular or of known form, the forms of perspective projection are for the most part characteristic for the depth of the object. If we look at houses, or other results of man's artistic activity, we know at the outset that the forms are for the most part plane surfaces at right angles to each other, with occasional circular or even spheroidal surfaces. And in fact, when we know so much, a correct perspective drawing is sufficient to produce the whole shape of the body. This is also the case with the figures of men and animals which are familiar to us, and whose forms moreover show two symmetrical halves. The best perspective drawing is however of but little avail in the case of irregular shapes, rough blocks of rock and ice, masses of foliage, and the like; that this is so, is best seen in photographs, where the perspective and shading may be absolutely correct, and yet the total impression is indistinct and confused.

When human habitations are seen in a picture, they represent to the observer the direction of the horizontal surfaces at the place at which they stand; and in comparison therewith the inclination of the ground, which without them would often be difficult to represent.

The apparent magnitude which objects, whose actual magnitude is known, present in different parts of the picture must also be taken into account. Men and animals, as well as familiar trees, are useful to the painter in this respect. In the more distant centre of the landscape they appear smaller than in the foreground, and thus their apparent magnitude furnishes a measure of the distance at which they are placed.

Shadows, and more especially double ones, are of great importance. You all know how much more distinct is the impression which a well-shaded drawing gives as distinguished from an outline; the shading is hence one of the most difficult, but at the same time most effective, elements in the productions of the draughtsman and painter. It is his task to imitate the fine gradation and transitions of light and shade on rounded surfaces, which are his chief means of expressing their modelling, with all their fine changes of curvature; he must take into account the extension or restriction of the sources of light, and the mutual reflection of the surfaces on each other. While the modifications of the lighting on the surface of bodies themselves is often dubious—for instance, an intaglio of a medal may, with a particular illumination, produce the impression of reliefs which are only illuminated from the other side—double shadows, on the contrary, are undoubted indications that the body which throws the shadow is nearer the source of light than that which receives the shadow. This rule is so completely without exception, that even in stereoscopic views a falsely placed double shadow may destroy or confuse the entire illusion.

The various kinds of illumination are not all equally favourable for obtaining the full effect of shadows. When the observer looks at the objects in the same direction as that in which light falls upon them, he sees only their illuminated sides and nothing of the shadow; the whole relief which the shadows could give then disappears. If the object is between the source of light and the observer he only sees the shadows. Hence we need lateral illumination for a picturesque shading; and over surfaces which like those of plane or hilly land only present slightly moving figures, we require light which is almost in the direction of the surface itself, for only such a one gives shadows. This is one of the reasons which makes illumination by the rising or the setting sun so effective. The forms of the landscape become more distinct. To this must also be added the influence of colour, and of aerial light, which we shall subsequently discuss.

Direct illumination from the sun, or from a flame, makes the shadows sharply defined, and hard. Illumination from a very wide luminous surface, such as a cloudy sky, makes them confused, or destroys them altogether. Between these two extremes there are transitions; illumination by a portion of the sky, defined by a window, or by trees, &c., allows the shadows to be more or less prominent according to the nature of the object. You must have seen of what importance this is to photographers, who have to modify their light by all manner of screens and curtains in order to obtain well-modelled portraits.

Of more importance for the representation of depth than the elements hitherto enumerated, and which are more or less of local and accidental significance, is what is called *aerial perspective*. By this we understand the optical action of the light, which the illuminated masses of air, between the observer and distant objects, give. This arises from a fine opacity in the atmosphere, which never entirely disappears. If, in a transparent medium, there are fine transparent particles of varying density and varying refrangibility, in so far as they are struck by it, they deflect the light passing through such a medium, partly by reflection and partly by refraction; to use an optical expression, they *scatter* it in all directions. If the opaque particles are sparsely distributed, so that a great part of the light can pass through them without being deflected, distant objects are seen in sharp, well-defined outlines through such a medium, while at the same time a portion of the light which is deflected is distributed in the transparent medium as an opaque halo. Water rendered turbid by a few drops of milk shows this dispersion of the light and cloudiness very distinctly. The light in this case is deflected by the microscopic globules of butter which are suspended in the milk.

In the ordinary air of our rooms, this turbidity is very apparent when the room is closed, and a ray of sunlight is admitted through a narrow aperture. We see then some of these solar particles, large enough to be distinguished by the naked eye, while others form a fine homogeneous turbidity. But even the latter must consist mainly of suspended particles of organic substances, for, according to an observation of Tyndall, they can be burnt. If the flame of a spirit lamp is placed directly below the path of these rays, the air rising from the flame stands out quite dark in the surrounding bright turbidity; that is to say, the air rising from the flame has been quite freed from dust. In the open air, besides dust and occasional smoke, we must often also take into account the turbidity arising from incipient aqueous deposits, where the temperature of moist air sinks so far that the water retained in it can no longer exist as invisible vapour. Part of the water settles then in the form of fine drops, as a kind of the very finest aqueous dust, and forms a finer or denser fog; that is to say, cloud. The turbidity which forms in hot sunshine and dry air may arise, partly from dust which the ascending currents of warm air whirl about; and partly from the irregular mixture of cold and warm layers of air of different density, as is seen in the tremulous motion of the lower layers of air over surfaces irradiated by the sun. But science can as yet give no explanation of the turbidity in the higher regions of the atmosphere which produces the blue of the sky; we do not know whether it arises from suspended particles of foreign substances, or whether the molecules of air themselves may not act as turbid particles in the luminous ether.

The colour of the light reflected by the opaque particles mainly depends on their magnitude. When a block of wood floats on water, and by a succession of falling drops we produce small wave-rings near it, these are repelled by the floating wood as if it were a solid wall. But in the long waves of the sea, a block of wood would be rocked about without the waves being thereby materially disturbed in their progress. Now light is well known to be an undulatory motion of the ether which fills all space. The red and yellow rays have the longest waves, the blue and violet the shortest. Very fine particles, therefore, which disturb the uniformity of the ether, will accordingly reflect the latter rays more markedly than the red and yellow rays. The light of turbid media is bluer, the finer are the opaque particles; while the larger particles of uniform light reflect all colours, and therefore give a whitish turbidity. Of this kind is the celestial blue, that is, the colour of the turbid atmosphere as seen against dark cosmical space. The purer and the more transparent the air, the bluer is the sky. In like manner it is bluer and darker when we ascend high mountains, partly because the air at great heights is freer from turbidity, and partly because there is less air above us. But the same blue, which is seen against the dark celestial space, also occurs against dark terrestrial objects; for instance, when a thick layer of illuminated air is between us and masses of deeply shaded or wooded hills. The same aerial light makes the sky blue, as well as the mountains; excepting that in the former case it is pure, while in the latter it is mixed with the light from objects behind; and moreover it belongs to the coarser turbidity of the lower regions of the atmosphere, so that it is whiter. In hot countries, and with dry air, the aerial turbidity is also finer in the lower regions of the air, and therefore the blue in front of distant terrestrial objects is more like that of the sky. The clearness and the pure colours of Italian landscapes depend mainly on this fact. On high mountains, particularly in the morning, the aerial turbidity is often so slight that the colours of the most distant objects can scarcely be distinguished from those of the nearest. The sky may then appear almost bluish-black.

Conversely, the denser turbidity consists mainly of coarser particles, and is therefore whitish. As a rule, this is the case in the lower layers of air, and in states of weather in which the aqueous vapour in the air is near its point of condensation.

On the other hand, the light which reaches the eye of the observer after having passed through a long layer of air, has been robbed of part of its violet and blue by scattered reflections; it therefore appears yellowish to reddish-yellow or red, the former when the turbidity is fine, the latter when it is coarse. Thus the sun and the moon at their rising and setting, and also distant brightly illuminated mountain-tops, especially snow-mountains, appear coloured.

These colourations are moreover not peculiar to the air, but occur in all cases in which a transparent substance is made turbid by the admixture of another transparent substance. We see it, as we have observed, in diluted milk, and in water to which a few drops of eau de Cologne have been added, whereby the ethereal oils and resins dissolved by the latter, separate out and produce the turbidity. Excessively fine blue clouds, bluer even than the air, may be produced, as Tyndall has observed, when the sun's light is allowed to exert its decomposing action on the vapours of certain carbon compounds. Goethe called attention to the universality of this phenomenon, and endeavoured to base upon it his theory of colour.

By aerial perspective we understand the artistic representation of aerial turbidity; for the greater or less predominance of the aerial colour above the colour of the objects, shows their varying distance very definitely; and landscapes more especially acquire the appearance of depth. According to the weather, the turbidity of the air may be greater or less, more white or more blue. Very clear air, as sometimes met with after continued rain, makes the distant mountains appear small and near; whereas, when the air contains more vapour, they appear large and distant.

This latter is decidedly better for the landscape painter, and the high transparent landscapes of mountainous regions, which so often lead the Alpine climber to under-estimate the distance and the magnitude of the mountain-tops before him, are also difficult to turn to account in a picturesque manner. Views from the valleys, and from seas and plains in which the aerial light is faintly but markedly developed, are far better; not only do they allow the various distances and magnitudes of what is seen to stand out, but they are on the other hand favourable to the artistic unity of colouration.

Although aerial colour is most distinct in the greater depths of landscape, it is not entirely wanting in front of the near objects of a room. What is seen to be isolated and well defined, when sunlight passes into a dark room through a hole in the shutter, is also not quite wanting when the whole room is lighted. Here, also, the aerial lighting must stand out against the background, and must somewhat deaden the colours in comparison with those of nearer objects; and these differences, also, although far more delicate than against the background of a landscape, are important for the historical, genre, or portrait painter; and when they are carefully observed and imitated, they greatly heighten the distinctness of his representation.

II. Shade.

The circumstances which we have hitherto discussed indicate a profound difference, and one which is exceedingly important for the perception of solid form, between the visual image which our eyes give, when we stand before objects, and that which the picture gives. The choice of the objects to be represented in pictures is thereby at once much restricted. Artists are well aware that there is much which cannot be represented by the means at their disposal. Part of their artistic skill consists in the fact that by a suitable grouping, position, and turn of the objects, by a suitable choice of the point of view, and by the mode of lighting, they learn to overcome the unfavourable conditions which are imposed on them in this respect.

It might at first sight appear that of the requisite truth to nature of a picture, so much would remain that, seen from the proper point of view, it would at least produce the same distribution of light, colour, and shadow in its field of view, and would produce in the interior of the eye exactly the same image on the retina as the object represented would do if we had it actually before us, and looked at it from a definite,

ON THE RELATION OF OPTICS TO PAINTING. 95

fixed point of view. It might seem to be an object of pictorial skill to aim at producing, under the given limitations, the *same* effect as is produced by the object itself.

If we proceed to examine whether, and how far, painting can satisfy such a condition, we come upon difficulties before which we should perhaps shrink, if we did not know that they had been already overcome.

Let us begin with the simplest case; with the quantitative relations between luminous intensities. If the artist is to imitate exactly the impression which the object produces on our eye, he ought to be able to dispose of brightness and darkness equal to that which nature offers. But of this there can be no idea. Let me give a case in point. Let there be, in a picture-gallery, a desert-scene, in which a procession of Bedouins, shrouded in white, and of dark negroes, marches under the burning sunshine; close to it a bluish moonlight scene, where the moon is reflected in the water, and groups of trees, and human forms, are seen to be faintly indicated in the darkness. You know from experience that both pictures, if they are well done, can produce with surprising vividness the representation of their objects; and yet, in both pictures, the brightest parts are produced with the same white-lead, which is but slightly altered by admixtures; while the darkest parts are produced with the same black. Both, being hung on the same wall, share the same light, and the brightest as well as the darkest parts of the two scarcely differ as concerns the degree of their brightness.

How is it, however, with the actual degrees of brightness represented? The relation between the brightness of the sun's light, and that of the moon, was measured by Wollaston, who compared their intensities with that of the light of candles of the same material. He thus found that the luminosity of the sun is 800,000 times that of the brightest light of a full moon.

An opaque body, which is lighted from any source whatever, can, even in the most favourable case, only emit as much light as falls upon it. Yet, from Lambert's observations, even the whitest bodies only reflect about two fifths of the incident light. The sun's rays, which proceed parallel from the sun, whose diameter is 85,000 miles, when they reach us, are distributed uniformly over a sphere 195 millions of miles in diameter. Its density and illuminating power is here only the one forty-thousandth of that with which it left the sun's surface; and Lambert's number leads to the conclusion that even the brightest white surface on which the sun's rays fall vertically, has only the one hundred-thousandth part of the brightness of the sun's disk. The moon however is a gray body, whose mean brightness is only about one fifth of that of the purest white.

And when the moon irradiates a body of the purest white on the earth, its brightness is only the hundred-thousandth part of the brightness of the moon itself; hence the sun's disk is 80,000 million times brighter than a white which is irradiated by the full moon.

Now pictures which hang in a room are not lighted by the direct light of the sun, but by that which is reflected from the sky and clouds. I do not know of any direct measurements of the ordinary brightness of the light in a picture gallery, but estimates may be made from known data. With strong upper light and bright light from the clouds, the brightest white on a picture has probably 1-20th of the brightness of white directly lighted by the sun; it will generally be only 1-40th, or even less.

Hence the painter of the desert, even if he gives up the representation of the sun's disk, which is always very imperfect, will have to represent the glaringly lighted garments of his Bedouins with a white which, in the most favourable case, shows only the 1-20th part of the brightness which corresponds to actual fact. If he could bring it, with its lighting unchanged, into the desert near the white there, it would seem like a dark grey. I found in fact, by an experiment, that lamp-black, lighted by the sun, is not less than half as bright, as shaded white in the brighter part of a room.

On the picture of the moon, the same white which has been used for depicting the Bedouins' garments must be used for representing the moon's disk, and its reflection in the water; although the real moon has only one fifth of this brightness, and its reflection in water still less. Hence white garments in moonlight, or marble surfaces, even when the artist gives them a grey shade, will always be ten to twenty times as bright in his picture as they are in reality.

On the other hand, the darkest black which the artist could apply would be scarcely sufficient to represent the real illumination of a white object on which the moon shone. For even the deadest black coatings of lamp-black, black velvet, when powerfully lighted appear grey, as we often enough know to our cost, when we wish to shut off superfluous light. I investigated a coating of lamp-black, and found its brightness to be about $\frac{1}{100}$ that of white paper. The brightest colours of a painter are only about one hundred times as bright as his darkest shades.

The statements I have made may perhaps appear exaggerated. But they depend upon measurements, and you can control them by well-known observations. According to Wollaston, the light of the full moon is equal to that of a candle burning at a distance of 12 feet. You know that we cannot read by the light of the full moon, though we can read at a distance of three or four feet from a candle. Now assume that you suddenly passed from a room in daylight to a vault perfectly dark, with the exception of the light of a single candle. You would at first think you were in absolute darkness, and at most you would only recognise the candle itself. In any case, you would not recognise the slightest trace of any objects at a distance of 12 feet from the candle. These however are the objects whose illumination is the same as that which the moonlight gives. You would only become accustomed to the darkness after some time, and you would then find your way about without difficulty.

If, now, you return to the daylight, which before was perfectly comfortable, it will appear so dazzling that you will perhaps have to close the eyes, and only be able to gaze round with a painful glare. You see thus that we are concerned here not with minute, but with colossal, differences. How now is it possible that, under such circumstances, we can imagine there is any similarity between the picture and reality?

Our discussion of what we did not see at first, but could afterwards see in the vault, points to the most important element in the solution; it is the varying extent to which our senses are deadened by light; a process to which we can attach the same name, fatigue, as that for the corresponding one in the muscle. Any activity of our nervous system diminishes its power for the time being. The muscle is tired by work, the brain is tired by thinking, and by mental operations; the eye is tired by light, and the more so the more powerful the light. Fatigue makes it dull and insensitive to new impressions, so that it appreciates strong ones only moderately, and weak ones not at all.

But now you see how different is the aim of the artist when these circumstances are taken into account. The eye of the traveller in the desert, who is looking at the caravan, has been dulled to the last degree by the dazzling sunshine; while that of the wanderer by moonlight has been raised to the extreme of sensitiveness. The condition of one who is looking at a picture differs from both the above cases by possessing a certain mean degree of sensitiveness. Accordingly, the painter must endeavour to produce by his colours, on the moderately sensitive eye of the spectator, the same impression as that which the desert, on the one hand, produces on the deadened, and the moonlight, on the other hand, creates on the untired eye of its observer. Hence, along with the actual luminous phenomena of the outer world, the different physiological conditions of the eye play a most important part in the work of the artist. What he has to give is not a mere transcript of the object, but a translation of his impression into another scale of sensitiveness, which belongs to a different degree of impressibility of the observing eye, in which the organ speaks a very different dialect in responding to the impressions of the outer world.

In order to understand to what conclusions this leads, I must first of all explain the law which Fechner discovered for the scale of sensitiveness of the eye, which is a particular case of the more general *psychophysical law* of the relations of the various sensuous impressions to the irritations which produce them. This law may be expressed as follows: *Within very wide limits of brightness, differences in the strength of light are equally distinct or appear equal in sensation, if they form an equal fraction of the total quantity of light compared.* Thus, for instance, differences in intensity of one hundredth of the total amount can be recognised without great trouble with very different strengths of light, without exhibiting material differences in the certainty and facility of the estimate, whether the brightest daylight or the light of a good candle be used.

The easiest method of producing accurately measurable differences in the brightness of two white surfaces, depends on the use of rapidly rotating disks. If a disk, like the adjacent one in Fig. 3, is made to rotate very rapidly (that is, 20 to 30 times in a second), it appears to the eye to be covered with three grey rings as in Fig. 4. The reader must, however, figure to himself the grey of these rings, as it appears on

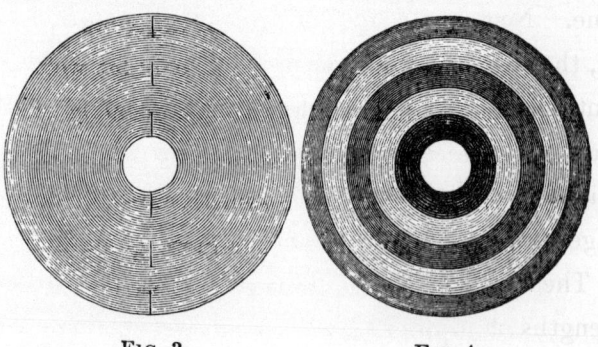

FIG. 3.   FIG. 4.

the rotating disk of Fig. 3, as a scarcely perceptible shade of the ground. When the rotation is rapid each ring of the disk appears illuminated, as if all the light which fell upon it had been uniformly distributed over its entire surface. Those rings, in which are the black bands, have somewhat less light than the quite white ones, and if the breadth of the marks is compared with the length of half the circumference of the corresponding ring, we get the fraction by which the intensity of the light in the white ground of the disk is diminished in the ring in question. If the bands are all equally broad, as in Fig. 3, the inner rings appear darker than the outer ones, for in this latter case the same loss of light is distributed over a larger area than in the former. In this way extremely delicate shades of brightness may be obtained, and by this method, when the strength of the illumination varies, the brightness always diminishes by the *same proportion* of its total value. Now it is found, in accordance with Fechner's law, that the distinctness of the rings is nearly constant for very different strengths of light. We exclude, of course, the cases of too dazzling or of too dim a light. In both cases the finer distinctions can no longer be perceived by the eye.

The case is quite different when for different strengths of illumination we produce differences which always correspond to the same quantity of light. If, for instance, we close the shutter of a room at daytime, so that it is quite dark, and now light it by a candle, we can discriminate without difficulty the shadows, such as that of the hand, thrown by the candle on a sheet of white paper. If, however, the shutters are again opened, so that daylight enters the room, for the same position of the hand we can no longer recognise the shadow, although there falls on that part of the white sheet, which is not struck by this shadow, the same excess of candle-light as upon the parts shaded by the hand. But this small quantity of light disappears in comparison with the newly added daylight, provided that this strikes all parts of the white sheet uniformly. You see then that, while the difference between candle-light and darkness can be easily perceived, the equally great difference between daylight, on the one hand, and day light plus candle-light on the other, can be no longer recognised.

This law is of great importance in discriminating between various degrees of brightness of natural objects. A white body appears white because it reflects a large fraction, and a grey body appears grey because it reflects a small fraction, of incident light. For different intensities of illumination, the difference of brightness between the two will always correspond to the same fraction of their total brightness, and hence will be equally perceptible to our eyes, provided we do not approach too near to the upper or the lower limit of the brightness, for which Fechner's law no longer holds. Hence, on the whole, the painter can produce what appears an equal difference for the spectator of his picture, notwithstanding the varying strength of light in the gallery, provided he gives to his colours the same *ratio* of brightness as that which actually exists.

For, in fact, in looking at natural objects, the absolute brightness in which they appear to the eye varies within very wide limits, according to the intensity of the light, and the sensitiveness of the eye. That which is constant is only the ratio of the brightness in which surfaces of various depth of colour appear to us when lighted to the same amount. But this ratio of brightness is for us the perception, from which we form our judgment as to the lighter or darker colour of the bodies we see. Now this ratio can be imitated by the painter without restraint, and in conformity with nature, to evoke in us the same conception as to the nature of the bodies seen. A truthful imitation in this respect would be attained within the limits in which Fechner's law holds, if the artist reproduced the fully lighted parts of the objects which he has to represent with pigments, which, with the same light, were equal to the colours to be represented. This is approximately the case. On the whole, the painter chooses coloured pigments which almost exactly reproduce the colours of the bodies represented, especially for objects of no great depth, such as portraits, and which are only darker in the shaded parts. Children begin to paint on this principle; they imitate one colour by another; and, in like manner also, nations in which painting has remained in a childish stage. Perfect artistic painting is only reached when we have succeeded in imitating the action of light upon the eye, and not merely the pigments; and only when we look at the object of pictorial representation from this point of view, will it be possible to understand the variations from nature which artists have to make in the choice of their scale of colour and of shade.

These are, in the first case, due to the circumstance that Fechner's law only holds for mean degrees of brightness; while, for a brightness which is too high or too low, appreciable divergences are met with.

At both extremes of luminous intensity the eye is less sensitive for differences in light than is required by that law. With a very strong light it is dazzled; that is, its internal activity cannot keep pace with the external excitations; the nerves are too soon tired. Very bright objects appear almost always to be equally bright, even when there are, in fact, material differences in their luminous intensity. The light at the edge of the sun is only about half as bright as that at the centre, yet none of you will have noticed that, if you have not looked through coloured glasses, which reduce the brightness to a convenient extent. With a weak light the eye is also less sensitive, but from the opposite reason. If a body is so feebly illuminated that we scarcely perceive it, we shall not be able to perceive that its brightness is lessened by a shadow by the one hundredth or even by a tenth.

It follows from this, that, with moderate illumination, darker objects become more like the darkest objects, while with greater illumination brighter objects become more like the brightest than should be the case in accordance with Fechner's law, which holds for mean degrees of illumination. From this results, what, for painting, is an extremely characteristic difference between the impression of very powerful and very feeble illumination.

When painters wish to represent glowing sunshine, they make all objects almost equally bright, and thus produce with their moderately bright colours the impression which the sun's glow makes upon the dazzled eye of the observer. If, on the contrary, they wish to represent moonshine, they only indicate the very brightest objects, particularly the reflection of moonlight on shining surfaces, and keep everything so dark as to be almost unrecognisable; that is to say, they make all dark objects more like the deepest dark which they can produce with their colours, than should be the case in accordance with the true ratio of the luminosities. In both cases they express, by their gradation of the lights, the *insensitiveness* of the eye for differences of too bright or too feeble lights. If they could employ the colour of the dazzling brightness of full sunshine, or of the actual dimness of moonlight, they would not need to represent the gradation of light in their picture other than it is in nature; the picture would then make the same impression on the eye as is produced by equal degrees of brightness of actual objects. The alteration in the scale of shade which has been described is necessary because the colours of the picture are seen in the mean brightness of a moderately lighted room, for which Fechner's law holds; and therewith objects are to be represented whose brightness is beyond the limits of this law.

We find that the older masters, and pre-eminently Rembrandt, employ the same deviation, which corresponds to that actually seen in moonlight landscapes; and this in cases in which it is by no means wished to produce the impression of moonshine, or of a similar feeble light. The brightest parts of the objects are given in these pictures in bright, luminous yellowish colours; but the shades towards the black are made very marked, so that the darker objects are almost lost in an impermeable darkness. But this darkness is covered with the yellowish haze of powerfully lighted aërial masses, so that, notwithstanding their darkness, these pictures give the impression of sunlight, and the very marked gradation of the shadows, the contours of the faces and figures, are made extremely prominent. The deviation from strict truth to nature is very remarkable in this shading, and yet these pictures give particularly bright and vivid aspects of the objects. Hence they are of particular interest for understanding the principles of pictorial illumination.

In order to explain these actions we must, I think, consider that while Fechner's law is approximately correct for those mean lights which are agreeable to the eye, the deviations which are so marked, for too high or too low lights, are not without some influence in the region of the middle lights. We have to observe more closely in order to perceive this influence. It is found, in fact, that when the very finest differences of shade are produced on a rotating disk, they are only visible by a light which about corresponds to the illumination of a white paper on a bright day, which is lighted by the light of the sky, but is not directly struck by the sun. With such a light, shades of $\frac{1}{150}$ or $\frac{1}{180}$ of the total intensity can be recognised. The light in which pictures are looked at is, on the contrary, much feebler; and if we are to retain the same distinctness of the finest shadows and of the modelling of the contours which it produces, the gradations of shade in the picture must be somewhat stronger than corresponds to the exact luminous intensities. The darkest objects of the picture thereby become unnaturally dark, which is however not detrimental to the object of the artist if the attention of the observer is to be directed to the brighter parts. The great artistic effectiveness of this manner shows us that the chief emphasis is to be laid on imitating difference of brightness and not on absolute brightness; and that the greatest differences in this latter respect can be borne without perceptible incongruity, if only their gradations are imitated with expression.

III. COLOUR.

With these divergences in brightness are connected certain divergences in colour, which, physiologically, are caused by the fact that the scale of sensitiveness is different for different colours. The strength of the sensation produced by light of a particular colour, and for a given intensity of light, depends altogether on the special reaction of that complex of nerves which are set in operation by the action of the light in question. Now all our sensations of colour are admixtures of three simple sensations; namely, of red, green, and violet,[1] which, by a not improbable supposition of Thomas Young, can be apprehended quite independently of each other by three different systems of nerve-fibres. To this independence of the different sensations of colour corresponds their independence in the gradation of intensity. Recent measurements[2] have shown that the sensitiveness of our eye for feeble shadows is greatest in the blue and least in the red. A difference of $\frac{1}{205}$ to $\frac{1}{268}$ of the intensity can be observed in the blue, and with an untired eye

---

[1] Helmholtz's *Popular Scientific Lectures*, pp. 232–52.
[2] Dobrowolsky in *Graefe's Archiv für Ophthalmologie*, vol. xviii. part i. pp. 24–92.

of $\frac{1}{16}$ in the red; or when the colour is dimmed by being looked at for a long time, a difference of $\frac{1}{50}$ to $\frac{1}{70}$.

Red therefore acts as a colour towards whose shades the eye is relatively less sensitive than towards that of blue. In agreement with this, the impression of glare, as the intensity increases, is feebler in red than in blue. According to an observation of Dove, if a blue and a red paper be chosen which appear of equal brightness under a mean degree of white light, as the light is made much dimmer the blue appears brighter, and as the light is much strengthened, the red. I myself have found that the same differences are seen, and even in a more striking manner, in the red and violet spectral colours, and, when their intensity is increased only moderately, by the same fraction for both.

Now the impression of white is made up of the impressions which the individual spectral colours make on our eye. If we increase the brightness of white, the strength of the sensation for the red and yellow rays will relatively be more increased than that for the blue and violet. In bright white, therefore, the former will produce a relatively stronger impression than the latter; in dull white the blue and bluish colours will have this effect. Very bright white appears therefore yellowish, and dull white appears bluish. In our ordinary way of looking at the objects about us, we are not so readily conscious of this; for the direct comparison of colours of very different shade is difficult, and we are accustomed to see in this alteration in the white the result of different illumination of one and the same white object, so that in judging pigment-colours we have learnt to eliminate the influence of brightness.

If however to the painter is put the problem of imitating, with faint colours, white irradiated by the sun, he can attain a high degree of resemblance; for by an admixture of yellow in his white he makes this colour preponderate just as it would preponderate in actual bright light, owing to the impression on the nerves. It is the same impression as that produced if we look at a clouded landscape through a yellow glass, and thereby give it the appearance of a sunny light. The artist will, on the contrary, give a bluish tint to moonlight, that is, a faint white; for the colours on the picture must, as we have seen, be far brighter than the colour to be represented. In moonshine scarcely any other colour can be recognised than blue; the blue starry sky or blue colours may still appear distinctly coloured, while yellow and red can only be seen as obscurations of the general bluish white or grey.

I will again remind you that these changes of

colour would not be necessary if the artist had at his disposal colours of the same brightness, or the same faintness, as are actually shown by the bodies irradiated by the sun or by the moon.

The change of colour, like the scale of shade, previously discussed, is a subjective action which the artist must represent objectively on his canvas, since moderately bright colours cannot produce them.

We observe something quite similar in regard to the phenomena of *Contrast*. By this term we understand cases in which the colour or brightness of a surface appears changed by the proximity of a mass of another colour or shade, and, in such a manner, that the original colour appears darker by the proximity of a brighter shade, and brighter by that of a darker shade; while by a colour of a different kind it tends towards the complementary tint.

The phenomena of contrast are very various, and depend on different causes. One class, *Chevreul's simultaneous Contrast*, is independent of the motions of the eyes, and occurs with surfaces where there are very slight differences in colour and shade. This contrast appears both on the picture and in actual objects, and is well known to painters. Their mixtures of colours on the palette often appear quite different to what they are on the picture. The changes of colour which are here met with are often very striking; I will not, however, enter upon them, for they produce no divergence between the picture and reality.

The second class of phenomena of contrast, and one which, for us, is more important, is met with in changes of direction of the glance, and more especially between surfaces in which there are great differences of shade and of colour. As the eye glides over bright and dark, or coloured objects and surfaces, the impression of each colour changes, for it is depicted on portions of the retina which directly before were struck by other colours and lights, and were therefore changed in their sensitiveness to an impression. This kind of contrast is therefore essentially dependent on movements of the eye, and has been called by Chevreul, 'successive Contrast.'

We have already seen that the retina is more sensitive in the dark to feeble light than it was before. By strong light, on the contrary, it is dulled, and is less sensitive to feeble lights which it had before perceived. This latter process is designated as 'Fatigue' of the retina; an exhaustion of the capability of the retina by its own activity, just as the muscles by their activity become tired.

I must here remark that the fatigue of the retina by light does not necessarily extend to the whole surface; but when only a small portion of this membrane is struck by a minute, defined picture it can also be locally developed in this part only.

You must all have observed the dark spots which move about in the field of vision, when we have been looking for only a short time towards the setting sun, and which physiologists call *negative after-images* of the sun. They are due to the fact that only those parts of the retina which are actually struck by the image of the sun in the eye, have become insensitive to a new impression of light. If, with an eye which is thus locally tired, we look towards a uniformly bright surface, such as the sky, the tired parts of the retina are more feebly and more darkly affected than the other portions, so that the observer thinks he sees dark spots in the sky, which move about with his sight. We have then in juxtaposition, in the bright parts of the sky, the impression which these make upon the untired parts of the retina, and in the dark spots their action on the tired portions. Objects, bright like the sun, produce negative after-images in the most striking manner; but with a little attention they may be seen even after much more moderate impressions of light. A longer time is required in order to develop such an impression, so that it may be distinctly recognised, and a definite point of the bright object must be fixed, without moving the eye, so that its image may be distinctly formed on the retina, and only a limited portion of the retina be excited and tired, just as in producing sharp photographic portraits, the object must be stationary during the time of exposure in order that its image may not be displaced on the sensitive plate. The after-image in the eye is, as it were, a photograph on the retina, which becomes visible owing to the altered sensitiveness towards fresh light, but only remains stationary for a short time; it is longer, the more powerful and durable was the action of light.

If the object viewed was coloured, for instance red paper, the after-image is of the complementary colour on a grey ground; in this case of a bluish green.[1] Rose-red paper, on the contrary, gives a pure green after-image, green a rose-red, blue a yellow, and yellow a blue. These phenomena show that in the retina partial fatigue is possible for the several colours. According to Thomas Young's hypothesis of the existence of three systems of fibres in the visual nerves,[2] of which one set perceives red whatever the kind of irritation, the second green, and the third violet, with green light, only those fibres of the retina which are sensitive to green are powerfully excited and tired.

[1] In order to see this kind of image as distinctly as possible, it is desirable to avoid all movements of the eye. On a large sheet of dark grey paper a small black cross is drawn, the centre of which is steadily viewed, and a quadrangular sheet of paper of that colour whose after-image is to be observed is slid from the side, so that one of its corners touches the cross. The sheet is allowed to remain for a minute or two, the cross being steadily viewed, and it is then drawn suddenly away, without relaxing the view. In place of the sheet removed the after-image appears then on the dark ground.

[2] See Helmholtz's *Popular Lectures*, first series, p. 250.

If this same part of the retina is afterwards illuminated with white light, the sensation of green is enfeebled, while that of red and violet is vivid and predominant; their sum gives the sensation of purple, which mixed with the unchanged white ground forms rose-red.

In the ordinary way of looking at light and coloured objects, we are not accustomed to fix continuously one and the same point; for following with the gaze the play of our attentiveness, we are always turning it to new parts of the object as they happen to interest us. This way of looking, in which the eye is continually moving, and therefore the retinal image is also shifting about on the retina, has moreover the advantage of avoiding disturbances of sight, which powerful and continuous after-images would bring with them. Yet here also, after-images are not wanting; only they are shadowy in their contours, and of very short duration.

If a red surface be laid upon a grey ground, and if we look from the red over the edge towards the grey, the edges of the grey will seem as if struck by such an after-image of red, and will seem to be of a faint bluish green. But as the after-image rapidly disappears, it is mostly only those parts of the grey, which are nearest the red, which show the change in a marked degree.

This also is a phenomenon which is produced more strongly by bright light and brilliant saturated colours than by fainter light and duller colours. The artist, however, works for the most part with the latter. He produces most of his tints by mixture; each mixed pigment is, however, greyer and duller than the pure colour of which it is mixed, and even the few pigments of a highly saturated shade, which oil-painting can employ, are comparatively dark. The pigments employed in water-colours and coloured chalks are again comparatively white. Hence such bright contrasts, as are observed in strongly coloured and strongly lighted objects in nature, cannot be expected from their representation in the picture. If, therefore, with the pigments at his command, the artist wishes to reproduce the impression which objects give, as strikingly as possible, he must paint the contrasts which they produce. If the colours on the picture are as brilliant and luminous as in the actual objects, the contrasts in the former case would produce themselves as spontaneously as in the latter. Here, also, subjective phenomena of the eye must be objectively introduced into the picture, because the scale of colour and of brightness is different upon the latter.

With a little attention you will see that painters and draughtsmen generally make a plain uniformly lighted surface brighter, where it is close to a dark object, and darker, where it is near a light object. You will find that uniform grey surfaces are given a yellowish tint at the edge where there is a back-

ON THE RELATION OF OPTICS TO PAINTING. 119

ground of blue, and a rose-red tint where they impinge on green, provided that none of the light collected from the blue or green can fall upon the grey. Where the sun's rays passing through the green leafy shade of trees strike against the ground, they appear to the eye, tired with looking at the predominant green, of a rose-red tint; the whole daylight, entering through a slit, appears blue, compared with reddish yellow candle-light. In this way they are represented by the painter, since the colours of his pictures are not bright enough to produce the contrast without such help.

To the series of subjective phenomena, which artists are compelled to represent objectively in their pictures, must be associated certain phenomena of *irradiation.* By this is understood cases in which any brignt object in the field spreads its light or colour over the neighbourhood. The phenomena are the more marked the brighter is the radiating object, and the halo is brightest in the immediate neighbourhood of the bright object, but diminishes at a greater distance. These phenomena of irradiation are most striking around a very bright light on a dark ground. If the view of the flame itself is closed by a narrow dark object such as the finger, a bright misty halo disappears, which covers the whole neighbourhood, and, at the same time, any objects there may be in the dark part of the field of view are seen more distinctly. If the flame is partly screened by a ruler, this appears jagged where the flame projects beyond it. The luminosity in the neighbourhood of the flame is so intense, that its brightness can scarcely be distinguished from that of the flame itself; as is the case with all bright objects, the flame appears magnified, and as if spreading over towards the adjacent dark objects.

The cause of this phenomenon is quite similar to that of aërial perspective. It is due to a diffusion of light which arises from the passage of light through dull media, excepting that for the phenomena of aërial perspective the turbidity is to be sought in the air in front of the eye, while for true phenomena of irradiation it is to be sought in the transparent media of the eye. When even the healthiest human eye is examined by powerful light, the best being a pencil of sunlight concentrated on the side by a condensing lens, it is seen that the sclerotica and crystalline lens are not perfectly clear. If strongly illuminated, they both appear whitish and as if rendered turbid by a fine mist. Both are, in fact, tissues of fibrous structure, and are not therefore so homogeneous as a pure liquid or a pure crystal. Every inequality, however small, in the structure of a transparent body can, however, reflect some of the incident light— that is, can diffuse it in all directions.[1]

[1] I disregard here the view that irradiation in the eye depends on

The phenomena of irradiation also occur with moderate degrees of brightness. A dark aperture in a sheet of paper illuminated by the sun, or a small dark object on a coloured glass plate which is held against the clear sky, appear as if the colour of the adjacent surface were diffused over them.

Hence the phenomena of irradiation are very similar to those which produce the opacity of the air. The only essential difference lies in this, that the opacity by luminous air is stronger before distant objects which have a greater mass of air in front of them than before near ones; while irradiation in the eye sheds its halo uniformly over near and over distant objects.

Irradiation also belongs to the subjective phenomena of the eye which the artist represents objectively, because painted lights and painted sunlight are not bright enough to produce a distinct irradiation in the eye of the observer.

The representation which the painter has to give of the lights and colours of his object I have described as a translation, and I have urged that, as a general rule, it cannot give a copy true in all its details. The altered scale of brightness which the artist must apply in many cases is opposed to this. It is not the colours of the objects, but the impression which they

a diffusion of the excitation in the substance of the nerves, as this appears to me too hypothetical. Moreover, we are here concerned with the phenomena and not with their cause.

have given, or would give, which is to be imitated, so as to produce as distinct and vivid a conception as possible of those objects. As the painter must change the scale of light and colour in which he executes his picture, he only alters something which is subject to manifold change according to the lighting, and the degree of fatigue of the eye. He retains the more essential, that is, the *gradations* of brightness and tint. Here present themselves a series of phenomena which are occasioned by the manner in which the eye replies to an external irritation; and since they depend upon the intensity of this irritation they are not directly produced by the varied luminous intensity and colours of the picture. These objective phenomena, which occur on looking at the object, would be wanting if the painter did not represent them objectively on his canvas. The fact that they are represented is particularly significant for the kind of problem which is to be solved by a pictorial representation.

Now, in all translations, the individuality of the translator plays a part. In artistic productions many important points are left to the choice of the artist, which he can decide according to his individual taste, or according to the requirements of his subject. Within certain limits he can freely select the absolute brightness of his colours, as well as the strength of the shadows. Like Rembrandt, he may exaggerate them in order to obtain strong relief; or he may diminish them, with Fra Angelico and his modern imitators, in order to soften earthly shadows in the representation of sacred objects. Like the Dutch school, he may represent the varying light of the atmosphere, now bright and sunny, and now pale, or warm and cold, and thereby evoke in the observer moods which depend on the illumination and on the state of the weather; or by means of undisturbed air he may cause his figures to stand out objectively clear as it were, and uninfluenced by subjective impressions. By this means, great variety is attained in what artists call 'style' or 'treatment,' and indeed in their purely pictorial elements.

IV. Harmony of Colour.

We here naturally raise the question: If, owing to the small quantity of light and saturation of his colours, the artist seeks, in all kinds of indirect ways, by imitating subjective impressions to attain resemblance to nature, as close as possible, but still imperfect, would it not be more convenient to seek for means of obviating these evils? Such there are indeed. Frescoes are sometimes viewed in direct sunlight; transparencies and paintings on glass can utilise far higher degrees of brightness, and far more saturated colours; in dioramas and in theatrical decorations we may employ powerful artificial light, and, if need be, the electric light. But when I enumerate these branches of art, it will at once strike you that those works which we admire as the greatest masterpieces of painting, do not belong to this class; but by far the larger number of the great works of art are executed with the comparatively dull water or oil-colours, or at any rate for rooms with softened light. If higher artistic effects could be attained with colours lighted by the sun, we should undoubtedly have pictures which took advantage of this. Fresco painting would have led to this; or the experiments of Münich's celebrated optician Steinheil, which he made as a matter of science, that is, to produce oil paintings which should be looked at in bright sunshine, would not be isolated.

Experiment seems therefore to teach, that moderation of light and of colours in pictures is ever advantageous, and we need only look at frescoes in direct sunlight, such as those of the new Pinakothek in Münich, to learn in what this advantage consists. Their brightness is so great that we cannot look at them steadily for any length of time. And what in this case is so painful and so tiring to the eye, would also operate in a smaller degree if, in a picture, brilliant colours were used, even locally and to a moderate extent, which were intended to represent bright sunlight, and a mass of light shed over the picture. It is much easier to produce an accurate imitation of the feeble light of moonshine with artificial light in dioramas and theatre decorations.

We may therefore designate truth to Nature of a beautiful picture as an ennobled fidelity to Nature. Such a picture reproduces all that is essential in the impression, and attains full vividness of conception, but without injury or tiring the eye by the nude lights of reality. The differences between Art and Nature are chiefly confined, as we have already seen, to those matters which we can in reality only estimate in an uncertain manner, such as the absolute intensities of light.

That which is pleasant to the senses, the beneficial but not exhausting fatigue of our nerves, the feeling of comfort, corresponds in this case, as in others, to those conditions which are most favourable for perceiving the outer world, and which admit of the finest discrimination and observation.

It has been mentioned above that the discrimination of the finest shadows, and of the modelling which they express, is the most delicate under a certain mean brightness. I should like to direct your attention to another point which has great importance in painting: I refer to our natural delight in colours, which has undoubtedly a great influence upon our pleasure in the works of the painter. In its simplest expression, as pleasure in gaudy flowers, feathers, stones, in fireworks, and Bengal lights, this inclination has but little to do with man's sense of art; it only appears as the natural pleasure of the perceptive organism in the varying and multifarious excitation of its various nerves, which is necessary for its healthy continuance and productivity. But the thorough fitness in the construction of living organisms, whatever their origin, excludes the possibility that in the majority of healthy individuals an instinct should be developed or maintain itself which did not serve some definite purpose.

We have not far to seek for the delight in light and in colours, and for the dread of darkness; this coincides with the endeavour to see and to recognise surrounding objects. Darkness owes the greater part of the terror which it inspires to the fright of what is unknown and cannot be recognised. A coloured picture gives a far more accurate, richer, and easier conception than a similarly executed drawing, which only retains the contrasts of light and shade. A picture retains the latter, but has in addition the material for discrimination which colours afford; by which surfaces which appear equally bright in the drawing, owing to their different colour, are now assigned to various objects, or again as alike in colour are seen to be parts of the same, or of similar objects. In utilising the relations thus naturally given, the artist, by means of prominent colours, can direct and enchain the attention of the observer upon the chief objects of the picture; and by the variety of the garments he can discriminate the figures from each other, but complete each individual one in itself. Even the natural pleasure in pure, strongly saturated colours, finds its justification in this direction. The case is analogous to that in music, with the full, pure, well-sounding tones of a beautiful voice. Such a one is more expressive; that is, even the smallest change of its pitch, or its quality—any slight interruption, any tremulousness, any rising or falling in it—is at once more distinctly recognised by the hearer than could be the case with a less regular sound; and it seems also that the powerful excitation which it produces in the ear of the listener, arouses trains of ideas and passions more strongly than does a feebler excitation of the same kind. A pure, fundamental colour bears to small admixtures the same relation as a dark ground on which the slightest shade of light is visible. Any of the ladies present will have known how sensitive clothes of uniform saturated shades are to dirt, in comparison with grey or greyish-brown materials. This also corresponds to the conclusions from Young's theory of colours. According to this theory, the perception of each of the three fundamental colours arises from the excitation of only one kind of sensitive fibres, while the two others are at rest; or at any rate are but feebly excited. A brilliant, pure colour produces a powerful stimulus, and yet, at the same time, a great degree of sensitiveness to the admixture of other colours, in those systems of nerve-fibres which are at rest. The modelling of a coloured surface mainly depends upon the reflection of light of other colours which falls upon them from without. It is more particularly when the material glistens that the reflections of the bright places are preferably of the colour of the incident light. In the depth of the folds, on the contrary, the coloured surface reflects against itself, and thereby makes its own colour more saturated. A white surface, on the contrary, of great brightness, produces a dazzling effect, and is thereby insensitive to slight degrees of shade. Strong colours thus, by the powerful irritation which they produce, can enchain the eye of the observer, and yet be expressive for the slightest change of modelling or of illumination; that is, they are expressive in the artistic sense.

If, on the other hand, we coat too large surfaces, they produce fatigue for the prominent colour, and a diminution in sensitiveness towards it. This colour then becomes more grey, and on all surfaces of a different colour the complementary tint appears, especially on grey or black surfaces. Hence therefore clothes, and more particularly curtains, which are of too bright a single colour, produce an unsatisfactory and fatiguing effect; the clothes have moreover the disadvantage for the wearer that they cover face and hands with the complementary colour. Blue produces yellow, violet gives greenish yellow, bright purple gives green, scarlet gives blue, and, conversely, yellow gives blue, etc. There is another circumstance which the artist has to consider, that colour is for him an important means of attracting the attention of the observer. To be able to do this he must be sparing in the use of the pure colours, otherwise they distract the attention, and the picture becomes glaring. It is necessary, on the other hand, to avoid a onesided fatigue of the eye by too prominent a colour. This is effected either by introducing the prominent colour to a moderate extent upon a dull, slightly coloured ground, or by the juxtaposition of variously saturated colours, which produce a certain equilibrium of irritation in the eye, and, by the contrast in their afterimages, strengthen and increase each other. A green surface on which the green after-image of a purple one falls, appears to be a far purer green than without such an after-image. By fatigue towards purple, that is towards red and violet, any admixture of these two colours in the green is enfeebled, while this itself produces its full effect. In this way the sensation of green is purified from any foreign admixture. Even the purest and most saturated green, which Nature shows in the prismatic spectrum, may thus acquire a higher degree of saturation. We find thus that the other pairs of complementary colours, which we have mentioned, make each other more brilliant by their contrast, while colours which are very similar are detrimental to each other, and acquire a grey tint.

These relations of the colours to each other have manifestly a great influence on the degree of pleasure which different combinations of colours afford. Two colours may, without injury, be juxtaposed, which indeed are so similar as to look like varieties of the same colour, produced by varying degrees of light and shade. Thus, upon scarlet the more shaded parts appear of a carmine, or on a straw-colour they appear of a golden yellow.

If we pass beyond these limits, we arrive at unpleasant combinations, such as carmine and orange, or orange and straw-yellow. The distance of the colours must then be increased, so as to create pleasing combinations once more. The complementary colours are those which are most distant from each other. When these are combined, such, for instance, as straw-colour and ultramarine, or verdigris and purple, they have something insipid but crude; perhaps because we are prepared to expect the second colour to appear as an after-image of the first, and it does not sufficiently appear to be a new and independent element in the compound. Hence, on the whole, combinations of those pairs are most pleasing in which the second colour of the complementary tint is near the first, though with a distinct difference. Thus, scarlet and greenish blue are complementary. The combination produced when the greenish blue is allowed to glide either into ultramarine, or yellowish green (sap green), is still more pleasing. In the latter case, the combination tends towards yellow, and in the former, towards rose-red. Still more satisfactory combinations are those of three tints which bring about equilibrium in the impression of colour, and, notwithstanding the great body of colour, avoid a onesided fatigue of the eye, without falling into the baldness of complementary tints. To this belongs the combination which the Venetian masters used so much—red, green, and violet; as well as Paul Veronese's purple, greenish blue, and yellow. The former triad corresponds approximately to the three fundamental colours, in so far as these can be produced by pigments; the latter gives the mixtures of each pair of fundamental colours. It is however to be observed, that it has not yet been possible to establish rules for the harmony of colours with the same precision and certainty as for the consonance of tones. On the contrary, a consideration of the facts shows that a number of accessory influences come into play,[1] when once the coloured surface is also to produce, either wholly or in part, a representation of natural objects or of solid forms, or even if it only offers a resemblance with the representation of a relief, of shaded and of non-shaded surfaces. It is moreover often difficult to establish, as a matter of fact, what are the colours which produce the harmonic impression. This is pre-eminently the case with

[1] Conf. E. Brücke, *Die Physiologie der Farben für die Zwecke der Kunstgewerbe*. Leipzig, 1866. W. v. Bezold, *Die Farbenlehre in Einblick auf Kunst und Kunstgewerbe*. Braunschweig, 1874.

pictures in which the aërial colour, the coloured reflection and shade, so variously alter the tint of each single coloured surface when it is not perfectly smooth, that it is hardly possible to give an indisputable determination of its tint. In such cases, moreover, the direct action of the colour upon the eye is only a subordinate means; for, on the other hand, the prominent colours and lights must also serve for directing the attention to the more important points of the representation. Compared with these more poetical and psychological elements of the representation, considerations as to the pleasing effect of the colours are thrown into the background. Only in the pure ornamentation on carpets, draperies, ribbons, or architectonic surfaces is there free scope for pure pleasure in the colours, and only there can it develop itself according to its own laws.

In pictures, too, there is not, as a general rule, perfect equilibrium between the various colours, but one of them preponderates to an extent which corresponds to the dominant light. This is occasioned, in the first case, by the truthful imitation of physical circumstances. If the illumination is rich in yellow light, yellow colours will appear brighter and more brilliant than blue ones; for yellow bodies are those which preferably reflect yellow light; while that of blue is only feebly reflected, and is mainly absorbed.

Before the shaded parts of blue bodies, the yellow aërial light produces its effect, and imparts to the blue more or less of a grey tint. The same thing happens in front of red and green, though to a less extent, so that, in their shadows, these colours merge into yellow. This also is closely in accordance with the æsthetic requirements of artistic unity of composition in colour. This is caused by the fact that the divergent colours show a relation to the predominant colour, and point to it most distinctly in their shades. Where this is wanting, the various colours are hard and crude; and, since each one calls attention to itself, they make a motley and disturbing impression; and, on the other hand, a cold one, for the appearance of a flood of light thrown over the objects is wanting.

We have a natural type of the harmony which a well-executed illumination of masses of air can produce in a picture, in the light of the setting sun, which throws over the poorest regions a flood of light and colour, and harmoniously brightens them. The natural reason for this increase of aërial illumination lies in the fact, that the lower and more opaque layers of air are in the direction of the sun, and therefore reflect more powerfully; while at the same time the yellowish red colour of the light which has passed through the atmosphere becomes more distinct as the length of path increases which it has to traverse, and that further, this coloration is more pronounced as the background falls into shadow.

In summing up once more these considerations, we have first seen what limitations are imposed on truth to Nature in artistic representation; how the painter links the principal means which nature furnishes of recognising depths in the field of view, namely binocular vision, which indeed is even turned against him, as it shows unmistakably the flatness of the picture; how therefore the painter must carefully select, partly the perspective arrangement of his subject, its position and its aspect, and partly the lighting and shading, in order to give us a directly intelligible image of its magnitude, its shape, and distance, and how a truthful representation of aërial light is one of the most important means of attaining the object.

We then saw that even the scale of luminous intensity, as met with in the objects, must be transformed in the picture to one differing sometimes by a hundredfold; how here, the colour of the object cannot be simply represented by the pigment; that indeed it is necessary to introduce important changes in the distribution of light and dark, of yellowish and of bluish tints.

The artist cannot transcribe Nature; he must

translate her; yet this translation may give us an impression in the highest degree distinct and forcible, not merely of the objects themselves, but even of the greatly altered intensities of light under which we view them. The altered scale is indeed in many cases advantageous, as it gets rid of everything which, in the actual objects, is too dazzling, and too fatiguing for the eye. Thus the imitation of Nature in the picture is at the same time an ennobling of the impression on the senses. In this respect we can often give ourselves up more calmly and continuously, to the consideration of a work of art, than to that of a real object. The work of art can produce those gradations of light, and those tints in which the modelling of the forms is most distinct and therefore most expressive. It can bring forward a fulness of vivid fervent colours, and by skilful contrast can retain the sensitiveness of the eye in advantageous equilibrium. It can fearlessly apply the entire energy of powerful sensuous impressions, and the feeling of delight associated therewith, to direct and enchain the attention; it can use their variety to heighten the direct understanding of what is represented, and yet keep the eye in a condition of excitation most favourable and agreeable for delicate sensuous impressions.

If, in these considerations, my having continually laid much weight on the lightest, finest, and most accurate sensuous intelligibility of artistic representation, may seem to many of you as a very subordinate point—a point which, if mentioned at all by writers on æsthetics, is treated as quite accessory—I think this is unjustly so. The sensuous distinctness is by no means a low or subordinate element in the action of works of art; its importance has forced itself the more strongly upon me the more I have sought to discover the physiological elements in their action.

What effect is to be produced by a work of art, using this word in its highest sense? It should excite and enchain our attention, arouse in us, in easy play, a host of slumbering conceptions and their corresponding feelings, and direct them towards a common object, so as to give a vivid perception of all the features of an ideal type, whose separate fragments lie scattered in our imagination and overgrown by the wild chaos of accident. It seems as if we can only refer the frequent preponderance, in the mind, of art over reality, to the fact that the latter mixes something foreign, disturbing, and even injurious; while art can collect all the elements for the desired impression, and allow them to act without restraint. The power of this impression will no doubt be greater the deeper, the finer, and the truer to nature is the sensuous impression which is to arouse the series of images and the effects connected therewith. It must act certainly, rapidly, unequivocably, and with accuracy if it is to produce a vivid and powerful impression. These essentially are the points which I have sought to comprehend under the name of intelligibility of the work of art.

Then the peculiarities of the painters' technique (*Technik*), to which physiological optical investigation have led us, are often closely connected with the highest problems of art. We may perhaps think that even the last secret of artistic beauty—that is, the wondrous pleasure which we feel in its presence—is essentially based on the feeling of an easy, harmonic, vivid stream of our conceptions, which, in spite of manifold changes, flow towards a common object, bring to light laws hitherto concealed, and allow us to gaze in the deepest depths of sensation of our own minds.

ON THE ORIGIN

OF THE

PLANETARY SYSTEM.

Lecture delivered in Heidelberg and in Cologne, in 1871.

It is my intention to bring a subject before you to-day which has been much discussed—that is, the hypothesis of Kant and Laplace as to the formation of the celestial bodies, and more especially of our planetary system. The choice of the subject needs no apology. In popular lectures, like the present, the hearers may reasonably expect from the lecturer, that he shall bring before them well-ascertained facts, and the complete results of investigation, and not unripe suppositions, hypotheses, or dreams.

Of all the subjects to which the thought and imagination of man could turn, the question as to the origin of the world has, since remote antiquity, been the favourite arena of the wildest speculation. Beneficent and malignant deities, giants, Kronos who devours his children, Niflheim, with the ice-giant Ymir, who is killed by the celestial Asas,[1] that out of him the world may be constructed—these are all figures which fill the cosmogonic systems of the more cultivated of the peoples. But the universality of the fact, that each people develops its own cosmogonies, and sometimes in great detail, is an expression of the interest, felt by all, in knowing what is our own origin, what is the ultimate beginning of the things about us. And with the question of the beginning is closely connected that of the end of all things; for that which may be formed, may also pass away. The question about the end of things is perhaps of greater practical interest than that of the beginning.

Now, I must premise that the theory which I intend to discuss to-day was first put forth by a man who is known as the most abstract of philosophical thinkers; the originator of transcendental idealism and of the Categorical Imperative, Immanuel Kant. The work in which he developed this, the *General Natural Philosophy and Theory of the Heavens*, is one of his first publications, having appeared in his thirty-first year. Looking at the writings of this first period of his scientific activity, which lasted to about his fortieth year, we find that they belong mostly to

[1] Cox's *Aryan Mythology*, vol. i. 372. Longmans.

Natural Philosophy, and are far in advance of their times with a number of the happiest ideas. His philosophical writings at this period are but few, and partly like his introductory lecture, directly originating in some adventitious circumstance; at the same time the matter they contain is comparatively without originality, and they are only important from a destructive and partially sarcastic criticism. It cannot be denied that the Kant of early life was a natural philosopher by instinct and by inclination; and that probably only the power of external circumstances, the want of the means necessary for independent scientific research, and the tone of thought prevalent at the time, kept him to philosophy, in which it was only much later that he produced anything original and important; for the *Kritik der reinen Vernunft* appeared in his fifty-seventh year. Even in the later periods of his life, between his great philosophical works, he wrote occasional memoirs on natural philosophy, and regularly delivered a course of lectures on physical geography. He was restricted in this to the scanty measure of knowledge and of appliances of his time, and of the out-of-the-way place where he lived; but with a large and intelligent mind he strove after such more general points of view as Alexander von Humboldt afterwards worked out. It is exactly an inversion of the historical connection, when Kant's name is occasionally misused, to recommend that natural philosophy shall leave the inductive method, by which it has become great, to revert to the windy speculations of a so-called 'deductive method.' No one would have attacked such a misuse, more energetically and more incisively, than Kant himself if he were still among us.

The same hypothesis as to the origin of our planetary system was advanced a second time, but apparently quite independently of Kant, by the most celebrated of French astronomers, Simon, Marquis de Laplace. It formed, as it were, the final conclusion of his work on the mechanism of our system, executed with such gigantic industry and great mathematical acuteness. You see from the names of these two men, whom we meet as experienced and tried leaders in our course, that in a view in which they both agree, we have not to deal with a mere random guess, but with a careful and well-considered attempt to deduce conclusions as to the unknown past from known conditions of the present time.

It is in the nature of the case, that a hypothesis as to the origin of the world which we inhabit, and which deals with things in the most distant past, cannot be verified by direct observation. It may, however, receive direct confirmation, if, in the progress of scientific knowledge, new facts accrue to those already known, and like them are explained on the hypothesis; and particularly if survivals of the processes, assumed to have taken place in the formation of the heavenly bodies, can be proved to exist in the present.

Such direct confirmations of various kinds have, in fact, been formed for the view we are about to discuss, and have materially increased its probability.

Partly this fact, and partly the fact that the hypothesis in question has recently been mentioned in popular and scientific books, in connection with philosophical, ethical, and theological questions, have emboldened me to speak of it here. I intend not so much to tell you anything substantially new in reference to it, as to endeavour to give, as connectedly as possible, the reasons which have led to, and have confirmed it.

These apologies which I must premise, only apply to the fact that I treat a theme of this kind as a popular lecture. Science is not only entitled, but is indeed beholden, to make such an investigation. For her it is a definite and important question—the question, namely, as to the existence of limits to the validity of the laws of nature, which rule all that now surrounds us; the question whether they have always held in the past, and whether they will always hold in the future; or whether, on the supposition of an everlasting uniformity of natural laws, our conclusions from present circumstances as to the past, and as to the future, imperatively lead to an impossible state of things; that is, to the necessity of an infraction of natural laws, of a beginning which could not have been due to processes known to us. Hence, to begin such an investigation as to the possible or probable primeval history of our present world, is, considered as a question of science, no idle speculation, but a question as to the limits of its methods, and as to the extent to which existing laws are valid.

It may perhaps appear rash that we, restricted as we are, in the circle of our observations in space, by our position on this little earth, which is but as a grain of dust in our milky way; and limited in time by the short duration of the human race; that we should attempt to apply the laws which we have deduced from the confined circle of facts open to us, to the whole range of infinite space, and of time from everlasting to everlasting. But all our thought and our action, in the greatest as well as in the least, is based on our confidence in the unchangeable order of nature, and this confidence has hitherto been the more justified, the deeper we have penetrated into the interconnections of natural phenomena. And that the general laws, which we have found, also hold for the most distant vistas of space, has acquired strong actual confirmation during the past half-century.

In the front rank of all, then, is the law of gravitation. The celestial bodies, as you all know, float and move in infinite space. Compared with the enormous distances between them, each of us is but as a grain of dust. The nearest fixed stars, viewed even under the most powerful magnification, have no visible diameter; and we may be sure that even our sun, looked at from the nearest fixed stars, would only appear as a single luminous point; seeing that the masses of those stars, in so far as they have been determined, have not been found to be materially different from that of the sun. But, notwithstanding these enormous distances, there is an invisible tie between them which connects them together, and brings them in mutual interdependence. This is the force of gravitation, with which all heavy masses attract each other. We know this force as gravity, when it is operative between an earthly body and the mass of our earth. The force which causes a body to fall to the ground is none other than that which continually compels the moon to accompany the earth in its path round the sun, and which keeps the earth itself from fleeing off into space, away from the sun.

You may realise, by means of a simple mechanical model, the course of planetary motion. Fasten to the branch of a tree, at a sufficient height, or to a rigid bar, fixed horizontally in the wall, a silk cord, and at its end a small heavy body—for instance, a lead ball. If you allow this to hang at rest, it stretches the thread. This is the position of equilibrium of the ball. To indicate this, and keep it visible, put in the place of the ball any other solid body—for instance, a large terrestrial globe on a stand. For this purpose the ball must be pushed aside, but it presses against the globe, and, if taken away, it still tends to come back to it, because gravity impels it towards its position of equilibrium, which is in the centre of the sphere. And upon whatever side it is drawn, the same thing always happens. This force, which drives the ball towards the globe, represents in our model the attraction which the earth exerts on the moon, or the sun on the planets. After you have convinced yourselves of the accuracy of these facts, try to give the ball, when it is a little away from the globe, a slight throw in a lateral direction. If you have accurately hit the strength of the throw, the small ball will move round the large one in a circular path, and may retain this motion for some time; just as the moon persists in its course round the earth, or the planets about the sun. Now, in our model, the circles described by the lead ball will be continually narrower, because the opposing forces, the resistance of the air, the rigidity of the thread, friction, cannot be eliminated, in this case, as they are excluded in the planetary system.

If the path about the attracting centre is exactly circular, the attracting force always acts on the planets, or on the lead sphere, with equal strength. In this case, it is immaterial according to what law the force would increase or diminish at other distances from the centre in which the moving body does not come. If the original impulse has not been of the right strength in both cases, the paths will not

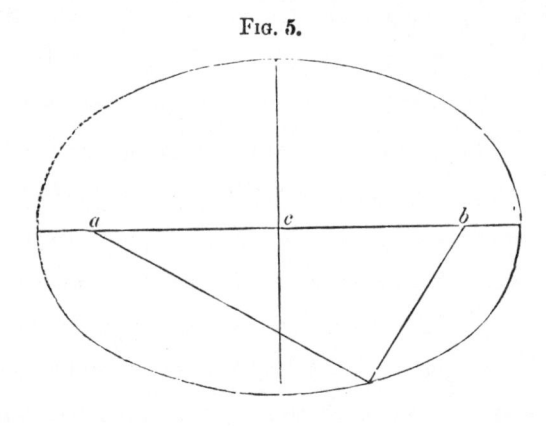

Fig. 5.

be circular but elliptical, of the form of the curved line in Fig. 5. But these ellipses lie in both cases differently as regards the attracting centre. In our model, the attracting force is stronger, the further the lead sphere is removed from its position of equilibrium. Under these circumstances, the ellipse of the path has such a position in reference to the attracting centre, that this is in the centre, c, of the ellipse. For planets, on the contrary, the attracting force is feebler the further it is removed from the attracting body, and this is the reason that an ellipse is described, one of whose foci lies in the centre of attraction. The two foci, a and b, are two points which lie symmetrically towards the ends of the ellipse, and are characterised by the property that the sum of their distances, $am + bm$, is the same from any given points.

Kepler had found that the paths of the planets are ellipses of this kind; and since, as the above example shows, the form and position of the orbit depend on the law according to which the magnitude of the attracting force alters, Newton could deduce from the form of the planetary orbits the well-known law of the force of gravitation, which attracts the planets to the sun, according to which this force decreases with increase of distance as the square of that distance. Terrestrial gravity must obey this law, and Newton had the wonderful self-denial to refrain from publishing his important discovery until it had acquired a direct confirmation; this followed from the observations, that the force which attracts the moon towards the earth, bears towards the gravity of a terrestrial body the ratio required by the above law.

In the course of the eighteenth century the power of mathematical analysis, and the methods of astronomical observation, increased so far that all the complicated actions, which take place between all the planets, and all their satellites, in consequence of the mutual action of each upon each, and which astronomers call disturbances—disturbance, that is to say, of the simpler elliptical motions about the sun, which each one would produce if the others were absent—that all these could be theoretically predicted from Newton's law, and be accurately compared with what actually takes place in the heavens. The development of this theory of planetary motion in detail was, as has been said, the merit of Laplace. The agreement between this theory, which was developed from the simple law of gravitation, and the extremely complicated and manifold phenomena which follow therefrom, was so complete and so accurate, as had never previously been attained in any other branch of human knowledge. Emboldened by this agreement, the next step was to conclude that where slight defects were still constantly found, unknown causes must be at work. Thus, from Bessel's calculation of the discrepancy between the actual and the calculated motion of Uranus, it was inferred that there must be another planet. The position of this planet was calculated by Leverrier and Adams, and thus Neptune, the most distant of all known at that time, was discovered.

But it was not merely in the region of the attraction of our sun that the law of gravitation was found to hold. With regard to the fixed stars, it was found that double stars moved about each other in elliptical paths, and that therefore the same law of gravitation must hold for them as for our planetary system. The distance of some of them could be calculated. The nearest of them, a, in the constellation of the Centaur, is 270,000 times further from the sun than the earth. Light, which has a velocity of 186,000 miles a second, which traverses the distance from the sun to the earth in eight minutes, would take four years to travel from a Centauri to us. The more delicate methods of modern astronomy have made it possible to determine distances which light would take thirty-five years to traverse; as, for instance, the Pole Star; but the law of gravitation is seen to hold, ruling the motion of the double stars, at distances in the heavens, which all the means we possess have hitherto utterly failed to measure.

The knowledge of the law of gravitation has here also led to the discovery of new bodies, as in the case of Neptune. Peters of Altona found, confirming therein a conjecture of Bessel, that Sirius, the most brilliant of the fixed stars, moves in an elliptical path about an invisible centre. This must have been due to an unseen companion, and when the excellent and powerful telescope of the University of Cambridge, in the United States, had been set up, this was discovered. It is not quite dark, but its light is so feeble that it can only be seen by the most perfect instruments. The mass of Sirius is found to be 13·76, and that of its satellite 6·71, times the mass of the sun; their mutual distance is equal to thirty-seven times the radius of the earth's orbit, and is therefore somewhat larger than the distance of Neptune from the sun.

Another fixed star, Procyon, is in the same case as Sirius, but its satellite has not yet been discovered.

You thus see that in gravitation we have discovered a property common to all matter, which is not confined to bodies in our system, but extends, as far in the celestial space, as our means of observation have hitherto been able to penetrate.

But not merely is this universal property of all mass shared by the most distant celestial bodies, as well as by terrestrial ones; but spectrum analysis has taught us that a number of well-known terrestrial elements are met with in the atmospheres of the fixed stars, and even of the nebulæ.

You all know that a fine bright line of light, seen through a glass prism, appears as a coloured band, red and yellow at one edge, blue and violet at the other, and green in the middle. Such a coloured image is called a spectrum—the rainbow is such a one, produced by the refraction of light, though not exactly by a prism; and it exhibits therefore the series of colours into which white sunlight can thus be decomposed.

The formation of the prismatic spectrum depends on the fact that the sun's light, and that of most ignited bodies, is made up of various kinds of light, which appear of different colours to our eyes, and the rays of which are separated from each other when refracted by a prism.

Now if a solid or a liquid is heated to such an extent that it becomes incandescent, the spectrum which its light gives is, like the rainbow, a broad coloured band without any breaks, with the well-known series of colours, red, yellow, green, blue, and violet, and in no wise characteristic of the nature of the body which emits the light.

The case is different if the light is emitted by an ignited gas, or by an ignited vapour—that is, a substance vaporised by heat. The spectrum of such a body consists, then, of one or more, and sometimes even a great number, of entirely distinct bright lines, whose position and arrangement in the spectrum is characteristic for the substances of which the gas or vapour consists, so that it can be ascertained, by means of spectrum analysis, what is the chemical constitution of the ignited gaseous body. Gaseous spectra of this kind are shown in the heavenly space by many nebulæ; for the most part they are spectra which show the bright line of ignited hydrogen and oxygen, and along with it a line which, as yet, has never been again found in the spectrum of any terrestrial element. Apart from the proof of two well-known terrestrial elements, this discovery was of the utmost importance, since it furnished the first unmistakable proof that the cosmical nebulæ are not, for the most part, small heaps of fine stars, but that the greater part of the light which they emit is really due to gaseous bodies.

The gaseous spectra present a different appearance when the gas is in front of an ignited solid whose temperature is far higher than that of the gas. The observer sees then a continuous spectrum of a solid, but traversed by fine dark lines, which are just visible in the places in which the gas alone, seen in front of a dark background, would show bright lines. The solar spectrum is of this kind, and also that of a great number of fixed stars. The dark lines of the solar spectrum, originally discovered by Wollaston, were first investigated and measured by Fraunhofer, and are hence known as Fraunhofer's lines.

Far more powerful apparatus was afterwards used by Kirchhoff, and then by Angström, to push the decomposition of light as far as possible. Fig. 6 represents an apparatus with four prisms, constructed by Steinheil for Kirchhoff. At the further end of the telescope B is a screen with a fine slit, representing a fine slice of light, which can be narrowed or widened by the small screw, and by which the light

154 ON THE ORIGIN OF THE PLANETARY SYSTEM.

under investigation can be allowed to enter. It then passes through the telescope B, afterwards through the

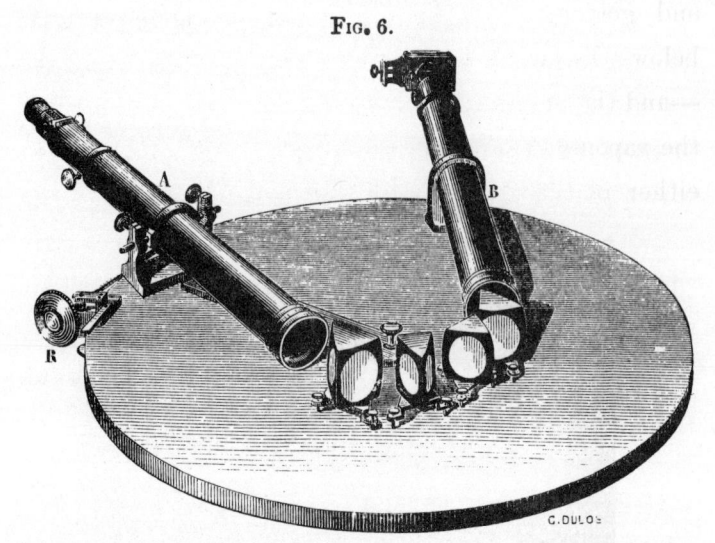

FIG. 6.

four prisms, and finally through the telescope A, from which it reaches the eye of the observer. Figs. 7, 8,

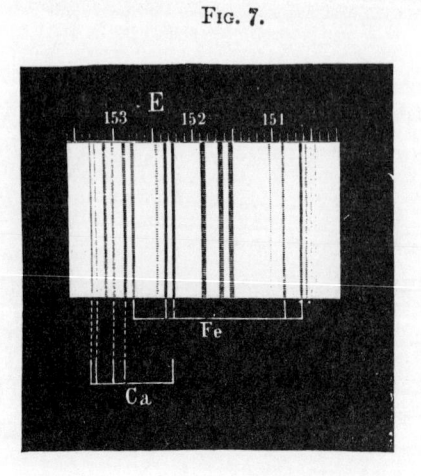

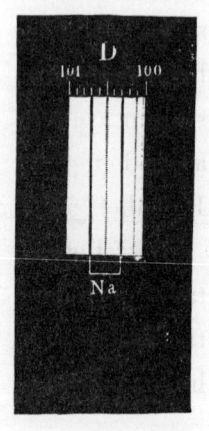

FIG. 7.

FIG. 8.

and 9 represent small portions of the solar spectrum as mapped by Kirchoff, taken from the green, yellow, and golden-yellow, in which the chemical symbols below—Fe (iron), Ca (calcium), Na (sodium), Pb (lead) —and the affixed lines, indicate the positions in which the vapours of these metals, when made incandescent, either in the flames or in the electrical spark, would

Fig. 9.

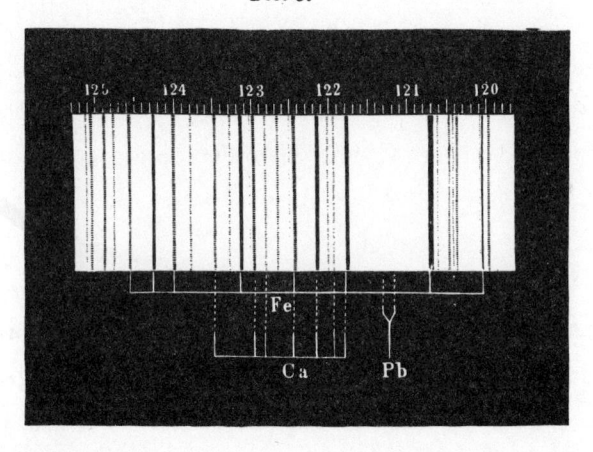

show bright lines. The numbers above them show how far these fractions of Kirchhoff's map of the whole system are apart from each other. Here, also, we see a predominance of iron lines. In the whole spectrum Kirchhoff found not less than 450.

It follows from this, that the solar atmosphere contains an abundance of the vapours of iron, which, by the way, justifies us in concluding what an enormously high temperature must prevail there. It shows, moreover, how our figs. 7, 8, and 9 indicate iron, calcium, and sodium, and also the presence of hydrogen, of zinc, of copper, and of the metals of magnesia, alumina, baryta, and other terrestrial elements. Lead, on the other hand, is wanting, as well as gold, silver, mercury, antimony, arsenic, and some others.

The spectra of several fixed stars are similarly constituted; they show systems of fine lines which can be identified with those of terrestrial elements. In the atmosphere of Aldebaran in Taurus there is, again, hydrogen, iron, magnesium, calcium, sodium, and also mercury, antimony, and bismuth; and, according to H. C. Vogel, there is in a Orionis the rare metal thallium; and so on.

We cannot, indeed, say that we have explained all spectra; many fixed stars exhibit peculiarly banded spectra, probably belonging to gases whose molecules have not been completely resolved into their atoms by the high temperature. In the spectrum of the sun, also, are many lines which we cannot identify with those of terrestrial elements. It is possible that they may be due to substances unknown to us, it is also possible that they are produced by the excessively high temperature of the sun, far transcending anything we can produce. But this is certain, that the known terrestrial substances are widely diffused in space, and especially nitrogen, which constitutes the greater part of our atmosphere, and hydrogen, an element in water, which indeed is formed by its combustion. Both have been found in the irresolvable nebulæ, and, from the inalterability of their shape, these must be masses of enormous dimensions and at an enormous distance. For this reason Sir W. Herschel considered that they did not belong to the system of our fixed stars, but were representatives of the manner in which other systems manifested themselves.

Spectrum analysis has further taught us more about the sun, by which he is brought nearer to us, as it were, than could formerly have seemed possible. You know that the sun is an enormous sphere, whose diameter is 112 times as great as that of the earth. We may consider what we see on its surface as a layer of incandescent vapour, which, to judge from the appearances of the sun-spots, has a depth of about 500 miles. This layer of vapour, which is continually radiating heat on the outside, and is certainly cooler than the inner masses of the sun, is, however, hotter than all our terrestrial flames—hotter even than the incandescent carbon points of the electrical arc, which represent the highest temperature attainable by terrestrial means. This can be deduced with certainty from Kirchhoff's law of the radiation of opaque bodies, from the greater luminous intensity of the sun. The older assumption, that the sun is a dark cool body, surrounded by a photosphere which only radiates heat and light externally, contains a physical impossibility.

Outside the opaque photosphere, the sun appears surrounded by a layer of transparent gases, which are hot enough to show in the spectrum bright coloured lines, and are hence called the *Chromosphere*. They show the bright lines of hydrogen, of sodium, of magnesium, and iron. In these layers of gas and of vapour about the sun enormous storms occur, which are as much greater than those of our earth in extent and in velocity as the sun is greater than the earth. Currents of ignited hydrogen burst out several thousands of miles high, like gigantic jets or tongues of flame, with clouds of smoke above them.[1] These structures could formerly only be viewed at the time of a total eclipse of the sun, forming what were called the rose-red protuberances. We now possess a method, devised by MM. Jansen and Lockyer, by which they may at any time be seen by the aid of the spectroscope.

On the other hand, there are individual darker parts on the sun's surface, what are called *sun-spots*, which were seen as long ago as by Galileo. They are funnel-shaped, the sides of the funnel are not so dark as the deepest part, the core. Fig. 10 represents such

[1] According to H. C. Vogel's observations in Bothkamp to a height of 70,000 miles. The spectroscopic displacement of the lines showed velocities of 18 to 23 miles in a second; and, according to Lockyer, of even 37 to 42 miles.

ON THE ORIGIN OF THE PLANETARY SYSTEM. 159

a spot according to Padre Secchi, as seen under powerful magnification. Their diameter is often more than many tens of thousands of miles, so that two or three earths could lie in one of them. These spots may stand for weeks or months, slowly changing, before

Fig. 10.

they are again resolved, and meanwhile several rotations of the sun may take place. Sometimes, however, there are very rapid changes in them. That the core is deeper than the edge of the surrounding penumbra follows from their respective displacements as they come near the edge, and are therefore seen in a very oblique direction. Fig. 11 represents in A to E the different aspects of such a spot as it comes near the edge of the sun.

Just on the edge of these spots there are spectroscopic indications of the most violent motion, and in their vicinity there are often large protuberances; they show comparatively often a rotatory motion. They may be considered to be places where the

FIG. 11.

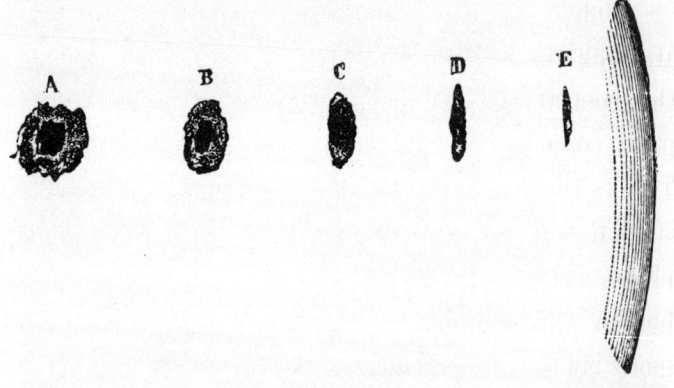

cooler gases from the outer layers of the sun's atmosphere sink down, and perhaps produce local superficial coolings of the sun's mass. To understand the origin of these phenomena, it must be remembered that the gases, as they rise from the hot body of the sun, are charged with vapours of difficultly volatile metals, which expand as they ascend, and partly by their expansion, and partly by radiation into space, must become cooled. At the same time, they deposit their more difficultly volatile constituents as fog or cloud. This cooling can only, of course, be regarded as comparative; their temperature is probably, even then, higher than any temperature attainable on the earth. If now the upper layers, freed from the heavier vapours, sink down, there will be a space over the sun's body which is free from cloud. They appear then as depressions, because about them are layers of ignited vapours as much as 500 miles in height.

Violent storms cannot fail to occur in the sun's atmosphere, because it is cooled on the outside, and the coolest and comparatively densest and heaviest parts come to lie over the hotter and lighter ones. This is the reason why we have frequent, and at times sudden and violent, movements in the earth's atmosphere, because this is heated from the ground made hot by the sun and is cooled above. With the far more colossal magnitude and temperature of the sun, its meteorological processes are on a far larger scale, and are far more violent.

We will now pass to the question of the permanence of the present condition of our system. For a long time the view was pretty generally held that, in its chief features at any rate, it was unchangeable. This opinion was based mainly on the conclusions at which Laplace had arrived as the final results of his long and laborious investigations, of the influence of planetary disturbances. By disturbances of the planetary motion astronomers understand, as I have already mentioned, those deviations from the purely elliptical motion which are due to the attraction of various planets and satellites upon each other. The attraction of the sun, as by far the largest body of our system, is indeed the chief and preponderating force which produces the motion of the planets. If it alone were operative, each of the planets would move continuously in a constant ellipse whose axes would retain the same direction and the same magnitude, making the revolutions always in the same length of time. But, in point of fact, in addition to the attraction of the sun there are the attractions of all other planets, which, though small, yet, in long periods of time, do effect slow changes in the plane, the direction, and the magnitude of the axes of its elliptical orbit. It has been asked whether these attractions in the orbit of the planet could go so far as to cause two adjacent planets to encounter each other, so that individual ones fall into the sun. Laplace was able to reply that this could not be the case; that all alterations in the planetary orbits produced by this kind of disturbance must periodically increase and decrease, and again revert to a mean condition. But it must not be forgotten that this result of Laplace's investigations only applies to disturbances due to the reciprocal attraction of planets upon each other, and on the assumption that no forces of other kinds have any influence on their motions.

On our earth we cannot produce such an everlasting motion as that of the planets seems to be; for resisting forces are continually being opposed to all movements of terrestrial bodies. The best known of these are what we call friction, resistance of the air, and inelastic impact.

Hence the fundamental law of mechanics, according to which every motion of a body on which no force acts goes on in a straight line for ever with unchanged velocity, never holds fully.

Even if we eliminate the influence of gravity in a ball, for example, which rolls on a plane surface, we see it go on for a while, and the further the smoother is the path; but at the same time we hear the rolling ball make a clattering sound—that is, it produces waves of sound in the surrounding bodies; there is friction even on the smoothest surface; this sets the surrounding air in vibration, and imparts to it some of its own motion. Thus it happens that its velocity is continually less and less until it finally ceases. In like manner, even the most carefully constructed wheel which plays upon fine points, once made to turn, goes on for a quarter of an hour, or even more, but then stops. For there is always some friction on the axles, and in addition there is the resistance of the air, which resistance is mainly due to that of the particles of air against each other, due to their friction against the wheel.

If we could once set a body in rotation, and keep it from falling, without its being supported by another body, and if we could transfer the whole arrangement to an absolute vacuum, it would continue to move for ever with undiminished velocity. This case, which cannot be realised on terrestrial bodies, is apparently met with in the planets with their satellites. They appear to move in the perfectly vacuous cosmical space, without contact with any body which could produce friction, and hence their motion seems to be one which never diminishes.

You see, however, that the justification of this conclusion depends on the question whether cosmical space is really quite vacuous. Is there nowhere any friction in the motion of the planets?

From the progress which the knowledge of nature has made since the time of Laplace, we must now answer both questions in the negative.

Celestial space is not absolutely vacuous. In the first place, it is filled by that continuous medium the agitation of which constitutes light and radiant heat, and which physicists know as the luminiferous ether. In the second place, large and small fragments of heavy matter, from the size of huge stones to that of dust, are still everywhere scattered; at any rate, in those parts of space which our earth traverses.

The existence of the luminiferous ether cannot be considered doubtful. That light and radiant heat are due to a motion which spreads in all directions has been sufficiently proved. For the transference of such a motion through space there must be something which can be moved. Indeed, from the magnitude of the action of this motion, or from that which the science of mechanics calls its *vis viva*, we may indeed assign certain limits for the density of this medium. Such a calculation has been made by Sir W. Thomson, the celebrated Glasgow physicist. He has found that the density may possibly be far less than that of the air in the most perfect exhaustion obtainable by a good air-pump; but that the mass of the ether cannot be absolutely equal to zero. A volume equal to that of the earth cannot contain less than 2,775 pounds of luminous ether.[1]

The phenomena in celestial space are in conformity with this. Just as a heavy stone flung through the air shows scarcely any influence of the resistance of the air, while a light feather is appreciably hindered; in like manner the medium which fills space is far too attenuated for any diminution to have been perceived

[1] This calculation would, however, lose its bases if Maxwell's hypothesis were confirmed, according to which light depends on electrical and magnetical oscillations

in the motion of the planets since the time in which we possess astronomical observations of their path. It is different with the smaller bodies of our system. Encke in particular has shown, with reference to the well-known small comet which bears his name, that it circulates round the sun in ever-diminishing orbits and in ever shorter periods of revolution. Its motion is similar to that of the circular pendulum which we have mentioned, and which, having its velocity gradually delayed by the resistance of the air, describes circles about its centre of attraction, which continually become smaller and smaller. The reason for this phenomenon is the following: The force which offers a resistance to the attraction of the sun on all comets and planets, and which prevents them from getting continually nearer to the sun, is what is called the centrifugal force—that is, the tendency to continue their motion in a straight line in the direction of their path. As the force of their motion diminishes, they yield by a corresponding amount to the attraction of the sun, and get nearer to it. If the resistance continues, they will continue to get nearer the sun until they fall into it. Encke's comet is no doubt in this condition. But the resistance whose presence in space is hereby indicated, must act, and has long continued to act, in the same manner on the far larger masses of the planets.

The presence of partly fine and partly coarse

heavy masses diffused in cosmical space is more distinctly revealed by the phenomena of asteroids and of meteorites. We know now that these are bodies which ranged about in cosmical space, before they came within the region of our terrestrial atmosphere. In the more strongly resisting medium which this atmosphere offers they are delayed in their motion, and at the same time are heated by the corresponding friction. Many of them may still find an escape from the terrestrial atmosphere, and continue their path through space with an altered and retarded motion. Others fall to the earth; the larger ones as meteorites, while the smaller ones are probably resolved into dust by the heat, and as such fall without being seen. According to Alexander Herschel's estimate, we may figure shooting-stars as being on an average of the same size as paving-stones. Their incandescence mostly occurs in the higher and most attenuated regions of the atmosphere, eighteen miles and more above the surface of the earth. As they move in space under the influence of the same laws as the planets and comets, they possess a planetary velocity of from eighteen to forty miles in a second. By this, also, we observe that they are in fact *stelle cadente*, falling stars, as they have long been called by poets.

This enormous velocity with which they enter our atmosphere is undoubtedly the cause of their becoming heated. You all know that friction heats the bodies rubbed. Every match that we ignite, every badly greased coach-wheel, every auger which we work in hard wood, teaches this. The air, like solid bodies, not only becomes heated by friction, but also by the work consumed in its compression. One of the most important results of modern physics, the actual proof of which is mainly due to the Englishman Joule, is that, in such a case, the heat developed is exactly proportional to the work expended. If, like the mechanicians, we measure the work done by the weight which would be necessary to produce it, multiplied by the height from which it must fall, Joule has shown that the work, produced by a given weight of water falling through a height of 425 metres, would be just sufficient to raise the same weight of water through one degree Centigrade. The equivalent in work of a velocity of eighteen to twenty-four miles in a second may be easily calculated from known mechanical laws; and this, transformed into heat, would be sufficient to raise the temperature of a piece of meteoric iron to 900,000 to 2,500,000 degrees Centigrade, provided that all the heat were retained by the iron, and did not, as it undoubtedly does, mainly pass into the air. This calculation shows, at any rate, that the velocity of the shooting-stars is perfectly adequate to raise them to the most violent incandescence. The temperatures attainable by terrestrial means scarcely exceed 2,000 degrees. In fact, the outer crusts of meteoric stones generally show traces of incipient fusion; and in cases in which observers examined with sufficient promptitude the stones which had fallen they found them hot on the surface, while the interior of detached pieces seemed to show the intense cold of cosmical space.

To the individual observer who casually looks towards the starry sky the meteorites appear as a rare and exceptional phenomenon. If, however, they are continuously observed, they are seen with tolerable regularity, especially towards morning, when they usually fall. But a single observer only views but a small part of the atmosphere; and if they are calculated for the entire surface of the earth it results that about seven and a half millions fall every day. In our regions of space, they are somewhat sparse and distant from each other. According to Alexander Herschel's estimates, each stone is, on an average, at a distance of 450 miles from its neighbours. But the earth moves through 18 miles every second, and has a diameter of 7,820 miles, and therefore sweeps through 876 millions of cubic miles of space every second, and carries with it whatever stones are contained therein.

Many groups are irregularly distributed in space, being probably those which have already undergone disturbances by planets. There are also denser swarms which move in regular elliptical orbits, cutting the earth's orbit in definite places, and therefore always occur on particular days of the year. Thus the 10th of August of each year is remarkable, and every thirty-three years the splendid fireworks of the 12th to the 14th of November repeats itself for a few years. It is remarkable that certain comets accompany the paths of these swarms, and give rise to the supposition that the comets gradually split up into meteoric swarms.

This is an important process. What the earth does is done by the other planets, and in a far higher degree by the sun, towards which all the smaller bodies of our system must fall; those, therefore, that are more subject to the influence of the resisting medium, and which must fall the more rapidly, the smaller they are. The earth and the planets have for millions of years been sweeping together the loose masses in space, and they hold fast what they have once attracted. But it follows from this that the earth and the planets were once smaller than they are now, and that more mass was diffused in space; and if we follow out this consideration it takes us back to a state of things in which, perhaps, all the mass now accumulated in the sun and in the planets, wandered loosely diffused in space. If we consider, further, that the small masses of meteorites as they now fall, have perhaps been formed by the gradual aggregation of fine dust, we see ourselves led to a primitive condition of fine nebulous masses.

From this point of view, that the fall of shooting-stars and of meteorites is perhaps only a small survival of a process which once built up worlds, it assumes far greater significance.

This would be a supposition of which we might admit the possibility, but which could not perhaps claim any great degree of probability, if we did not find that our predecessors, starting from quite different considerations, had arrived at the same hypothesis.

You know that a considerable number of planets rotate around the sun besides the eight larger ones, Mercury, Venus, the Earth, Mars, Jupiter, Saturn, Uranus, and Neptune; in the interval between Mars and Jupiter there circulate, as far as we know, 156 small planets or planetoids. Moons also rotate about the larger planets—that is, about the Earth and the four most distant ones, Jupiter, Saturn, Uranus, and Neptune; and lastly the Sun, and at any rate the larger planets, rotate about their own axes. Now, in the first place, it is remarkable that all the planes of rotation of the planets and of their satellites, as well as the equatorial planes of these planets, do not vary much from each other, and that in these planes all the rotation is in the same direction. The only considerable exceptions known are the moons of Uranus, whose plane is almost at right angles to the planes of the larger planets. It must at the same time be remarked that the coincidence, in the direction of these planes, is on the whole greater, the longer are the bodies and the larger the paths in question; while in the smaller bodies, and for the smaller paths, especially for the rotations of the planets about their own axes, considerable divergences occur. Thus the planes of all the planets, with the exception of Mercury and of the small ones between Mars and Jupiter, differ at most by three degrees from the path of the Earth. The equatorial plane of the Sun deviates by only seven and a half degrees, that of Jupiter only half as much. The equatorial plane of the Earth deviates, it is true, to the extent of twenty-three and a half degrees, and that of Mars by twenty-eight and a half degrees, and the separate paths of the small planet's satellites differ still more. But in these paths they all move direct, all in the same direction about the sun, and, as far as can be ascertained, also about their own axes, like the earth—that is, from west to east. If they had originated independently of each other, and had come together, any direction of the planes for each individual one would have been equally probable; a reverse direction of the orbit would have been just as probable as a direct one; decidedly elliptical paths would have been as probable as the almost circular

ON THE ORIGIN OF THE PLANETARY SYSTEM. 173

ones which we meet with in all the bodies we have named. There is, in fact, a complete irregularity in the comets and meteoric swarms, which we have much reason for considering to be formations which have only accidentally come within the sphere of the sun's attraction.

The number of coincidences in the orbits of the planets and their satellites is too great to be ascribed to accident. We must inquire for the reason of this coincidence, and this can only be sought in a primitive connection of the entire mass. Now, we are acquainted with forces and processes which condense an originally diffused mass, but none which could drive into space such large masses, as the planets, in the condition we now find them. Moreover, if they had become detached from the common mass, at a place much nearer the sun, they ought to have a markedly elliptical orbit. We must assume, accordingly, that this mass in its primitive condition extended at least to the orbit of the outermost planets.

These were the essential features of the considerations which led Kant and Laplace to their hypothesis. In their view our system was originally a chaotic ball of nebulous matter, of which originally, when it extended to the path of the most distant planet, many billions of cubic miles could contain scarcely a gramme of mass. This ball, when it had become detached from the nebulous balls of the adjacent fixed stars, possessed a slow movement of rotation. It became condensed under the influence of the reciprocal attraction of its parts; and, in the degree in which it condensed, the rotatory motion increased, and formed it into a flat disk. From time to time masses at the circumference of this disk became detached under the influence of the increasing centrifugal force; that which became detached formed again into a rotating nebulous mass, which either simply condensed and formed a planet, or during this condensation again repelled masses from the periphery, which became satellites, or in one case, that of Saturn, remained as a coherent ring. In another case, the mass which separated from the outside of the chief ball, divided into many parts, detached from each other, and furnished the swarms of small planets between Mars and Jupiter.

Our more recent experience as to the nature of star showers teaches us that this process of the condensation of loosely diffused masses to form larger bodies is by no means complete, but still goes on, though the traces are slight. The form in which it now appears is altered by the fact that meanwhile the gaseous or dust-like mass diffused in space had united under the influence of the force of attraction, and of the force of crystallisation of their constituents, to larger pieces than originally existed.

The showers of stars, as examples now taking place of the process which formed the heavenly bodies, are important from another point of view. They develop light and heat; and that directs us to a third series of considerations, which leads again to the same goal.

All life and all motion on our earth is, with few exceptions, kept up by a single force, that of the sun's rays, which bring to us light and heat. They warm the air of the hot zones, this becomes lighter and ascends, while the colder air flows towards the poles. Thus is formed the great circulation of the passage-winds. Local differences of temperature over land and sea, plains and mountains, disturb the uniformity of this great motion, and produce for us the capricious change of winds. Warm aqueous vapours ascend with the warm air, become condensed into clouds, and fall in the cooler zones, and upon the snowy tops of the mountains, as rain and as snow. The water collects in brooks, in rivers, moistens the plains, and makes life possible; crumbles the stones, carries their fragments along, and thus works at the geological transformation of the earth's surface. It is only under the influence of the sun's rays that the variegated covering of plants of the earth grows; and while they grow, they accumulate in their structure organic matter, which partly serves the whole animal kingdom as food, and serves man more particularly as fuel. Coals and lignites, the sources of power of our steam engines, are remains of primitive plants—the ancient production of the sun's rays.

Need we wonder if, to our forefathers of the Aryan race in India and Persia, the sun appeared as the fittest symbol of the Deity? They were right in regarding it as the giver of all life—as the ultimate source of almost all that has happened on earth.

But whence does the sun acquire this force? It radiates forth a more intense light than can be attained with any terrestrial means. It yields as much heat as if 1,500 pounds of coal were burned every hour upon each square foot of its surface. Of the heat which thus issues from it, the small fraction which enters our atmosphere furnishes a great mechanical force. Every steam-engine teaches us that heat can produce such force. The sun, in fact, drives on earth a kind of steam-engine whose performances are far greater than those of artificially constructed machines. The circulation of water in the atmosphere raises, as has been said, the water evaporated from the warm tropical seas to the mountain heights; it is, as it were, a water-raising engine of the most magnificent kind, with whose power no artificial machine can be even distantly compared. I have previously explained the mechanical equivalent of heat. Calculated by that standard, the work which the sun produces by its radiation is equal to the constant exertion of 7,000 horse-power for each square foot of the sun's surface.

For a long time experience had impressed on our mechanicians that a working force cannot be produced from nothing; that it can only be taken from the stores which nature possesses; which are strictly limited and which cannot be increased at pleasure—whether it be taken from the rushing water or from the wind; whether from the layers of coal, or from men and from animals, which cannot work without the consumption of food. Modern physics has attempted to prove the universality of this experience, to show that it applies to the great whole of all natural processes, and is independent of the special interests of man. These have been generalised and comprehended in the all-ruling natural law of the *Conservation of Force*. No natural process, and no series of natural processes, can be found, however manifold may be the changes which take place among them, by which a motive force can be continuously produced without a corresponding consumption. Just as the human race finds on earth but a limited supply of motive forces, capable of producing work, which it can utilise but not increase, so also must this be the case in the great whole of nature. The universe has its definite store of force, which works in it under ever varying forms; is indestructible, not to be increased, everlasting and unchangeable like matter itself. It seems as if Goethe had an idea of this when he makes the earth-spirit speak of himself as the representative of natural force.

> In the curren's of life, in the tempests of motion,
> In the fervour of art, in the fire, in the storm,
> Hither and thither,
> Over and under,
> Wend I and wander.
> Birth and the grave,
> Limitless ocean,
> Where the restless wave
> Undulates ever
> Under and over,
> Their seething strife
> Heaving and weaving
> The changes of life.
> At the whirling loom of time unawed,
> I work the living mantle of God.

Let us return to the special question which concerns us here: Whence does the sun derive this enormous store of force which it sends out?

On earth the processes of combustion are the most abundant source of heat. Does the sun's heat originate in a process of this kind? To this question we can reply with a complete and decided negative, for we now know that the sun contains the terrestrial elements with which we are acquainted. Let us select from among them the two, which, for the smallest mass, produce the greatest amount of heat when they combine; let us assume that the sun consists of hydrogen and oxygen, mixed in the proportion in which they would unite to form water. The mass of the sun is known, and also the quantity of heat produced by the union of known weights of oxygen and hydrogen. Calculation shows that under the above supposition, the heat resulting from their combustion would be sufficient to keep up the radiation of heat from the sun for 3,021 years. That, it is true, is a long time, but even profane history teaches that the sun has lighted and warmed us for 3,000 years, and geology puts it beyond doubt that this period must be extended to millions of years.

Known chemical forces are thus so completely inadequate, even on the most favourable assumption, to explain the production of heat which takes place in the sun, that we must quite drop this hypothesis.

We must seek for forces of far greater magnitude, and these we can only find in cosmical attraction. We have already seen that the comparatively small masses of shooting-stars and meteorites can produce extraordinarily large amounts of heat when their cosmical velocities are arrested by our atmosphere. Now the force which has produced these great velocities is gravitation. We know of this force as one acting on the surface of our planet when it appears as terrestrial gravity. We know that a weight *raised from the earth* can drive our clocks, and that in like manner the gravity of the water rushing down from the mountains works our mills.

If a weight falls from a height and strikes the ground its mass loses, indeed, the visible motion which it had as a whole—in fact, however, this motion is not lost; it is transferred to the smallest elementary particles of the mass, and this invisible vibration of the molecules is the motion of heat. Visible motion is transformed by impact, into the motion of heat.

That which holds in this respect for gravity, holds also for gravitation. A heavy mass, of whatever kind, which is suspended in space separated from another heavy mass, represents a force capable of work. For both masses attract each other, and, if unrestrained by centrifugal force, they move towards each other under the influence of this attraction; this takes place with ever-increasing velocity; and if this velocity is finally destroyed, whether this be suddenly, by collision, or gradually, by the friction of movable parts, it develops the corresponding quantity of the motion of heat, the amount of which can be calculated from the equivalence, previously established, between heat and mechanical work.

Now we may assume with great probability that very many more meteors fall upon the sun than upon the earth, and with greater velocity, too, and therefore give more heat. Yet the hypothesis, that the entire amount of the sun's heat which is continually lost by radiation, is made up by the fall of meteors, a hypothesis which was propounded by Mayer, and has been favourably adopted by several other physicists, is open, according to Sir W. Thomson's investigations, to objection; for, assuming it to hold, the mass of the sun should increase so rapidly that the consequences would have shown themselves in the accelerated motion of the planets. The entire loss of heat from the sun cannot at all events be produced in this way; at the most a portion, which, however, may not be inconsiderable.

If, now, there is no present manifestation of force sufficient to cover the expenditure of the sun's heat, the sun must originally have had a store of heat which it gradually gives out. But whence this store? We know that the cosmical forces alone could have produced it. And here the hypothesis, previously discussed as to the origin of the sun, comes to our aid. If the mass of the sun had been once diffused in cosmical space, and had then been condensed—that is, had fallen together under the influence of celestial gravity—if then the resultant motion had been destroyed by friction and impact, with the production of heat, the new world produced by such condensation must have acquired a store of heat not only of considerable, but even of colossal, magnitude.

Calculation shows that, assuming the thermal capacity of the sun to be the same as that of water, the temperature might be raised to 28,000,000 of degrees, if this quantity of heat could ever have been present in the sun at one time. This cannot be assumed, for such an increase of temperature would offer the greatest hindrance to condensation. It is probable rather that a great part of this heat, which was produced by condensation, began to radiate into space before this condensation was complete. But the heat which the sun could have previously developed by its condensation, would have been sufficient to cover its present expenditure for not less than 22,000,000 of years of the past.

And the sun is by no means so dense as it may become. Spectrum analysis demonstrates the presence of large masses of iron and of other known constituents of the rocks. The pressure which endeavours to condense the interior is about 800 times as great as that in the centre of the earth; and yet the density of the sun, owing probably to its enormous temperature, is less than a quarter of the mean density of the earth.

We may therefore assume with great probability that the sun will still continue in its condensation, even if it only attained the density of the earth—though it will probably become far denser in the interior owing to the enormous pressure—this would develop fresh quantities of heat, which would be sufficient to maintain for an additional 17,000,000 of years the same

ON THE ORIGIN OF THE PLANETARY SYSTEM.

intensity of sunshine as that which is now the source of all terrestrial life.

The smaller bodies of our system might become less hot than the sun, because the attraction of the fresh masses would be feebler. A body like the earth might, if even we put its thermal capacity as high as that of water, become heated to even 9,000 degrees, to more than our flames can produce. The smaller bodies must cool more rapidly as long as they are still liquid. The increase in temperature, with the depth, is shown in bore-holes and in mines. The existence of hot wells and of volcanic eruptions shows that in the interior of the earth there is a very high temperature, which can scarcely be anything than a remnant of the high temperature which prevailed at the time of its production. At any rate, the attempts to discover for the internal heat of the earth a more recent origin in chemical processes, have hitherto rested on very arbitrary assumptions; and, compared with the general uniform distribution of the internal heat, are somewhat insufficient.

On the other hand, considering the huge masses of Jupiter, of Saturn, of Uranus, and of Neptune, their small density, as well as that of the sun, is surprising, while the smaller planets and the moon approximate to the density of the earth. We are here reminded of the higher initial temperature, and the slower cooling, which characterises larger masses.[1] The moon, on the contrary, exhibits formations on its surface which are strikingly suggestive of volcanic craters, and point to a former state of ignition of our satellite. The mode of its rotation, moreover, that it always turns the same side towards the earth, is a peculiarity which might have been produced by the friction of a fluid. At present no trace of such a one can be perceived.

You see, thus, by what various paths we are constantly led to the same primitive conditions. The hypothesis of Kant and Laplace is seen to be one of the happiest ideas in science, which at first astounds us, and then connects us in all directions with other discoveries, by which the conclusions are confirmed until we have confidence in them. In this case another circumstance has contributed—that is, the observation that this process of transformation, which the theory in question presupposes, goes on still, though on a smaller scale, seeing

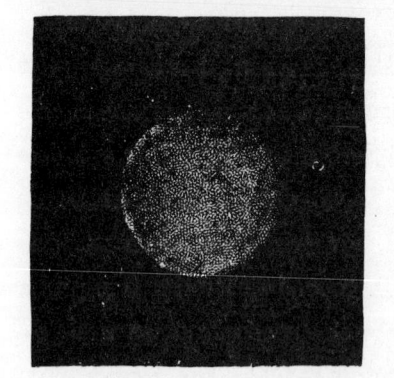

FIG. 12.

---

[1] Mr. Zoellner concludes from photometric measurements, which, however, need confirmation, that Jupiter still possesses a light of its own.

ON THE ORIGIN OF THE PLANETARY SYSTEM. 185

that all stages of that process can still be found to exist.

For as we have already seen, the larger bodies which are already formed go on increasing with the

Fig. 13.

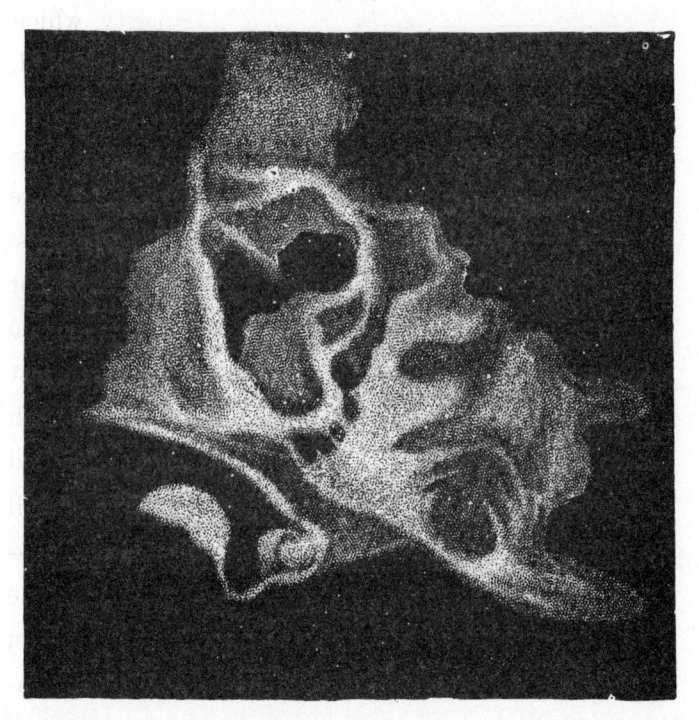

development of heat, by the attraction of the meteoric masses already diffused in space. Even now the smaller bodies are slowly drawn towards the sun by the resistance in space. We still find in the firmament of fixed stars, according to Sir J. Herschel's newest catalogue, over 5,000 nebulous spots, of which those whose light is sufficiently strong give for the most part a coloured spectrum of fine bright lines, as they appear in the spectra of the ignited gases. The nebulæ are partly rounded structures, which are called

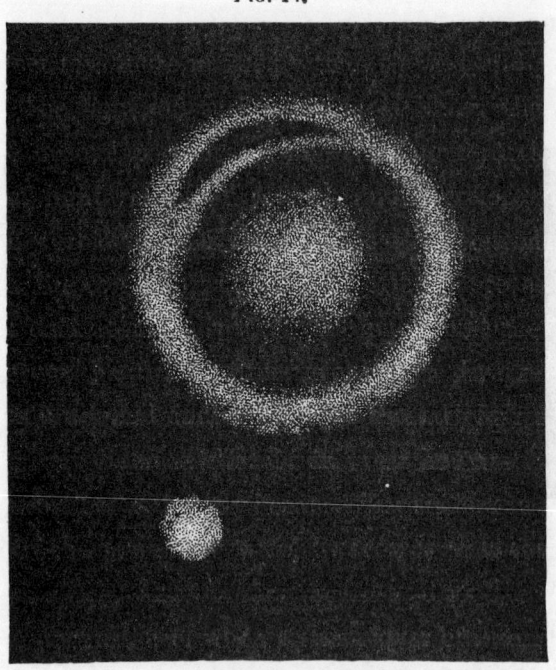

Fig. 14.

planetary nebulæ (fig. 12); sometimes wholly irregular in form, as the large nebula in Orion, represented in fig. 13; they are partly annular, as in the figures in fig. 14, from the Canes Venatici. They are for the most part feebly luminous over their whole surface, while the fixed stars only appear as luminous points.

ON THE ORIGIN OF THE PLANETARY SYSTEM 187

In many nebulæ small stars can be seen, as in figs. 15 and 16, from Sagittarius and Aurigo. More stars are continually being discovered in them, the better are the telescopes used in their analysis. Thus, before the discovery of spectrum analysis, Sir W. Herschel's former view might be regarded as the most probable, that that which we see to be nebulæ are only heaps of

Fig. 15. Fig. 16.

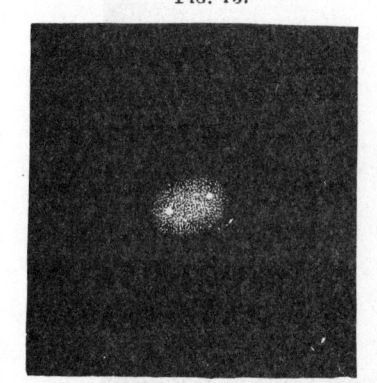

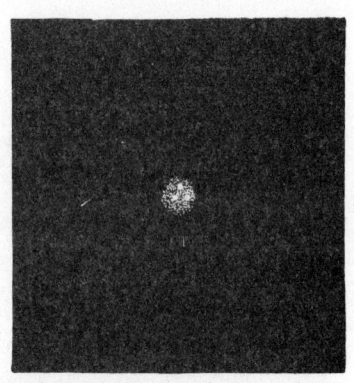

very fine stars, of other Milky Ways. Now, however, spectrum analysis has shown a gas spectrum in many nebulæ which contains stars, while actual heaps of stars show the continuous spectrum of ignited solid bodies. Nebulæ have in general three distinctly recognisable lines, one of which, in the blue, belongs to hydrogen, a second in bluish-green to nitrogen,[1] while the third, between the two, is of unknown origin. Fig. 17 shows

---

[1] Or perhaps also to oxygen. The line occurs in the spectrum of atmospheric air, and according to H. C. Vogel's observations was wanting in the spectrum of pure oxygen.

such a spectrum of a small but bright nebula in the Dragon. Traces of other bright lines are seen along with them, and sometimes also, as in fig. 17, traces of a continuous spectrum; all of which, however, are too feeble to admit of accurate investigation. It must be observed here that the light of very feeble objects which give a continuous spectrum are distributed by the spectroscope over a large surface, and are therefore greatly enfeebled or even extinguished, while the

FIG. 17.

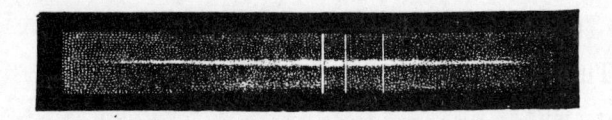

undecomposable light of bright gas lines remains undecomposed, and hence can still be seen. In any case, the decomposition of the light of the nebulæ shows that by far the greater part of their luminous surface is due to ignited gases, of which hydrogen forms a prominent constituent. In the planetary masses, the spherical or discoidal, it might be supposed that the gaseous mass had attained a condition of equilibrium; but most other nebulæ exhibit highly irregular forms, which by no means correspond to such a condition. As, however, their shape has either not at all altered, or not appreciably, since they have been known and observed, they must either have very little mass, or they must be of colossal size and distance. The former does not appear very probable, because small masses very soon give out their heat, and hence we are left to the second alternative, that they are of huge dimensions and distances. The same conclusion had been originally drawn by Sir W. Herschel, on the assumption that the nebulæ were heaps of stars.

With those nebulæ which, besides the lines of gases, also show the continuous spectrum of ignited denser bodies, are connected spots which are partly irresolvable and partly resolvable into heaps of stars, which only show the light of the latter kind.

The countless luminous stars of the heavenly firmament, whose number increases with each newer and more perfect telescope, associate themselves with this primitive condition of the worlds as they are formed. They are like our sun in magnitude, in luminosity, and on the whole also in the chemical condition of their surface, although there may be differences in the quantity of individual elements.

But we find also in space a third stadium, that of extinct suns; and for this also there are actual evidences. In the first place, there are, in the course of history, pretty frequent examples of the appearance of new stars. In 1572 Tycho Brahe observed such a one, which, though gradually burning paler, was visible for two years, stood still like a fixed star, and finally reverted to the darkness from which it had so suddenly emerged. The largest of them all seems to have been that observed by Kepler in the year 1604, which was brighter than a star of the first magnitude, and was observed from September 27, 1604, until March 1606. The reason of its luminosity was probably the collision with a smaller world. In a more recent case, in which on May 12, 1866, a small star of the tenth magnitude in the Corona suddenly burst out to one of the second magnitude, spectrum analysis showed that it was an outburst of ignited hydrogen which produced the light. This was only luminous for twelve days.

In other cases obscure heavenly bodies have discovered themselves by their attraction on adjacent bright stars, and the motions of the latter thereby produced. Such an influence is observed in Sirius and Procyon. By means of a new refracting telescope Messrs. Alvan Clarke and Pond, of Cambridge, U.S., have discovered in the case of Sirius a scarcely visible star, which has but little luminosity, but is almost seven times as heavy as the sun, has about half the mass of Sirius, and whose distance from Sirius is about equal to that of Neptune from the sun. The satellite of Procyon has not yet been seen; it appears to be quite dark.

Thus there are extinct suns. The fact that there are such lends new weight to the reasons which permit us to conclude that our sun also is a body which slowly gives out its store of heat, and thus will some time become extinct.

The term of 17,000,000 years which I have given may perhaps become considerably prolonged by the gradual abatement of radiation, by the new accretion of falling meteors, and by still greater condensation than that which I have assumed in that calculation. But we know of no natural process which could spare our sun the fate which has manifestly fallen upon other suns. This is a thought which we only reluctantly admit; it seems to us an insult to the beneficent Creative Power which we otherwise find at work in organisms and especially in living ones. But we must reconcile ourselves to the thought that, however we may consider ourselves to be the centre and final object of Creation, we are but as dust on the earth; which again is but a speck of dust in the immensity of space; and the previous duration of our race, even if we follow it far beyond our written history, into the era of the lake dwellings or of the mammoth, is but an instant compared with the primeval times of our planet; when living beings existed upon it, whose strange and unearthly remains still gaze at us from their ancient tombs; and far more does the duration of our race sink into insignificance compared with the enormous periods during which worlds have been in process of formation, and will still continue to form when our sun is extinguished, and our earth is either solidified in cold or is united with the ignited central body of our system.

But who knows whether the first living inhabitants of the warm sea on the young world, whom we ought perhaps to honour as our ancestors, would not have regarded our present cooler condition with as much horror as we look on a world without a sun? Considering the wonderful adaptability to the conditions of life which all organisms possess, who knows to what degree of perfection our posterity will have been developed in 17,000,000 of years, and whether our fossilised bones will not perhaps seem to them as monstrous as those of the Ichthyosaurus now do; and whether they, adjusted for a more sensitive state of equilibrum, will not consider the extremes of temperature, within which we now exist, to be just as violent and destructive as those of the older geological times appear to us? Yea, even if sun and earth should solidify and become motionless, who could say what new worlds would not be ready to develop life? Meteoric stones sometimes contain hydrocarbons; the light of the heads of comets exhibits a spectrum which is most like that of the electrical light in gases containing hydrogen and carbon. But carbon is the element, which is characteristic of organic compounds, from which living bodies are built up. Who knows whether these bodies, which everywhere swarm through space, do not scatter germs of life wherever there is a new world, which has become capable of giving a dwelling-place to organic bodies? And this life we might perhaps consider as allied to ours in its primitive germ, however different might be the form which it would assume in adapting itself to its new dwelling-place.

However this may be, that which most arouses our moral feelings at the thought of a future, though possibly very remote, cessation of all living creation on the earth, is more particularly the question whether all this life is not an aimless sport, which will ultimately fall a prey to destruction by brute force? Under the light of Darwin's great thought we begin to see that not only pleasure and joy, but also pain, struggle, and death, are the powerful means by which nature has built up her finer and more perfect forms of life. And we men know more particularly that in our intelligence, our civic order, and our morality we are living on the inheritance which our forefathers have gained for us, and that which we acquire in the same way, will in like manner ennoble the life of our posterity. Thus the individual, who works for the ideal objects of humanity, even if in a modest position, and in a limited sphere of activity, may bear without fear the thought that the thread of his own consciousness will one day break.

But even men of such free and large order of minds as Lessing and David Strauss could not reconcile themselves to the thought of a final destruction of the living race, and with it of all the fruits of all past generations.

As yet we know of no fact, which can be established by scientific observation, which would show that the finer and complex forms of vital motion could exist otherwise than in the dense material of organic life; that it can propagate itself as the sound-movement of a string can leave its originally narrow and fixed home and diffuse itself in the air, keeping all the time its pitch, and the most delicate shade of its colour-tint; and that, when it meets another string attuned to it, starts this again or excites a flame ready to sing to the same tone. The flame even, which, of all processes in inanimate nature, is the closest type of life, may become extinct, but the heat which it produces continues to exist—indestructible, imperishable, as an invisible motion, now agitating the molecules of ponderable matter, and then radiating into boundless space as the vibration of an ether. Even there it retains the characteristic peculiarities of its origin, and it reveals its history to the inquirer who questions it by the spectroscope. United afresh, these rays may ignite a new flame, and thus, as it were, acquire a new bodily existence

Just as the flame remains the same in appearance, and continues to exist with the same form and structure, although it draws every minute fresh combustible vapour, and fresh oxygen from the air, into the vortex of its ascending current; and just as the wave goes on in unaltered form, and is yet being reconstructed every moment from fresh particles of water, so also in the living being, it is not the definite mass of substance, which now constitutes the body, to which the continuance of the individual is attached. For the material of the body, like that of the flame, is subject to continuous and comparatively rapid change—a change the more rapid, the livelier the activity of the organs in question. Some constituents are renewed from day to day, some from month to month, and others only after years. That which continues to exist as a particular individual is like the flame and the wave—only the form of motion which continually attracts fresh matter into its vortex and expels the old. The observer with a deaf ear only recognises the vibration of sound as long as it is visible and can be felt, bound up with heavy matter. Are our senses, in reference to life, like the deaf ear in this respect?

ADDENDUM.

The sentences on page 193 gave rise to a controversial attack by Mr. J. C. F. Zoellner, in his book 'On the Nature of the Comets,' on Sir W. Thomson, on which I took occasion to express myself briefly in the preface to the second part of the German translation of the 'Handbook of Theoretical Physics,' by Thomson and Tait. I give here the passage in question:—

'I will mention here a further objection. It refers to the question as to the possibility that organic germs may occur in meteoric stones, and be conveyed to the celestial bodies which have been cooled. In his opening Address at the Meeting of the British Association in Edinburgh, in August 1871, Sir W. Thomson had described this as "not unscientific." Here also, if there is an error, I must confess that I also am a culprit. I had mentioned the same view as a possible mode of explaining the transmission of organisms through space, even a little before Sir W. Thomson, in a lecture delivered in the spring of the same year at Heidelberg and Cologne, but not published. I cannot object if anyone considers this hypothesis to be in a high, or even in the highest, degree improbable. But to me it seems a perfectly correct scientific procedure, that when all our attempts fail in producing organisms from inanimate matter, we may inquire whether life has ever originated at all or not, and whether its germs have not been transported from one world to another, and have developed themselves wherever they found a favourable soil.

'Mr. Zoellner's so-called physical objections are but of very small weight. He recalls the history of meteoric stone, and adds (p. xxvi.): "If, therefore, that meteoric stones covered with organisms had escaped with a whole skin in the smashup of its mother-body, and had not shared the general rise of temperature, it must necessarily have first passed through the atmosphere of the earth, before it could deliver itself of its organisms for the purpose of peopling the earth."

'Now, in the first place, we know from repeated observations that the larger meteoric stones only become heated in their outside layer during their fall through the atmosphere, while the interior is cold, or even very cold. Hence all germs which there might be in the crevices would be safe from combustion in the earth's atmosphere. But even those germs which were collected on the surface when they reached the highest and most attenuated layer of the atmosphere would long before have been blown away by the powerful draught of air, before the stone reached the denser parts of the gaseous mass, where the compression would be sufficient to produce an appreciable heat. And, on the other hand, as far as the impact of two bodies is concerned, as Thomson assumes, the first consequences would be powerful mechanical motions, and only in the degree in which this would be destroyed by friction would heat be produced. We do not know whether that would last for hours, for days, or for weeks. The fragments, which at the first moment were scattered with planetary velocity, might escape without any disengagement of heat. I consider it even not improbable, that a stone, or shower of stones, flying through the higher regions of the atmosphere of a celestial body, carries with it a mass of air which contains unburned germs.

'As I have already remarked I am not inclined to suggest that all these possibilities are probabilities. They are questions the existence and signification of which we must remember, in order that if the case arise they may be solved by actual observations or by conclusions therefrom.'

ON

THOUGHT IN MEDICINE.

An Address delivered August 2, 1877, on the Anniversary of the Foundation of the Institute for the Education of Army Surgeons.

It is now thirty-five years since, on the 2nd August, I stood on the rostrum in the Hall of this Institute, before another such audience as this, and read a paper on the operation of Venal Tumours. I was then a pupil of this Institution, and was just at the end of my studies. I had never seen a tumour cut, and the subject-matter of my lecture was merely compiled from books; but book knowledge played at that time a far wider and a far more influential part in medicine than we are at present disposed to assign to it. It was a period of fermentation, of the fight between learned tradition and the new spirit of natural science, which would have no more of tradition, but wished to depend upon individual experience. The authorities at that time judged more favourably of my Essay than I did myself, and I still possess the books which were awarded to me as the prize.

The recollections which crowd in upon me on this occasion have brought vividly before my mind a picture of the then condition of our science, of our endeavours and of our hopes, and have led me to compare the past state of things with that into which it has developed. Much indeed has been accomplished.

Although all that we hoped for has not been fulfilled, and many things have turned out differently from what we wished, yet we have gained much for which we could not have dared to hope. Just as the history of the world has made one of its few giant steps before our eyes, so also has our science; hence an old student, like myself, scarcely recognises the somewhat matronly aspect of Dame Medicine, when he accidentally comes again in relation to her, so vigorous and so capable of growth has she become in the fountain of youth of the Natural Sciences.

I may, perhaps, retain the impression of this antagonism, more freshly than those of my contemporaries whom I have the honour to see assembled before me; and who, having remained permanently connected with science and practice, have been less struck and less surprised by great changes, taking place as they do by slow steps. This must be my excuse for speaking to you about the metamorphosis which has taken place in medicine during this period, and with the results of whose development you are better acquainted than I am. I should like the impression of this development and of its causes not to be quite lost on the younger of my hearers. They have no special incentive for consulting the literature of that period; they would meet with principles which appear as if written in a lost tongue, so that it is by no means easy for us to transfer ourselves into the mode of thought of a period which is so far behind us. The course of development of medicine is an instructive lesson on the true principles of scientific inquiry, and the positive part of this lesson has, perhaps, in no previous time been so impressively taught as in the last generation.

The task falls to me, of teaching that branch of the natural sciences which has to make the widest generalisations, and has to discuss the meaning of fundamental ideas; and which has, on that account, been not unfitly termed Natural Philosophy by the English-speaking peoples. Hence it does not fall too far out of the range of my official duties and of my own studies, if I attempt to discourse here of the principles of scientific method, in reference to the sciences of experience.

As regards my acquaintance with the tone of thought of the older medicine, independently of the general obligation, incumbent on every educated physician, of understanding the literature of his science and the direction as well as the conditions of its progress, there was in my case a special incentive. In my first professorship at Königsberg, from the year 1849 to 1856, I had to lecture each winter on general pathology—that is, on that part of the subject which contains the general theoretical conceptions of the nature of disease, and of the principles of its treatment.

General pathology was regarded by our elders as the fairest blossom of medical science. But in fact, that which formed its essence possesses only historical interest for the disciples of modern natural science.

Many of my predecessors have broken a lance for the scientific defence of this essence, and more especially Henle and Lotz. The latter, whose starting-point was also medicine, had, in his general pathology and therapeutics, arranged it very thoroughly and methodically and with great critical acumen.

My own original inclination was towards physics; external circumstances compelled me to commence the study of medicine, which was made possible to me by the liberal arrangements of this Institution. It had, however, been the custom of a former time to combine the study of medicine with that of the Natural Sciences, and whatever in this was compulsory I must consider fortunate; not merely that I entered medicine at a time in which any one who was even moderately at home in physical considerations found a fruitful virgin soil for cultivation; but I consider the study of medicine to have been that training which preached more impressively and more convincingly than any other could have done, the everlasting principles of all scientific work; principles which are so simple and yet are ever forgotten again; so clear and yet always hidden by a deceptive veil.

Perhaps only he can appreciate the immense importance and the fearful practical scope of the problems of medical theory, who has watched the fading eye of approaching death, and witnessed the distracted grief of affection, and who has asked himself the solemn questions, Has all been done which could be done to ward off the dread event? Have all the resources and all the means which Science has accumulated become exhausted?

Provided that he remains undisturbed in his study, the purely theoretical inquirer may smile with calm contempt when, for a time, vanity and conceit seek to swell themselves in science and stir up a commotion. Or he may consider ancient prejudices to be interesting and pardonable, as remains of poetic romance, or of youthful enthusiasm. To one who has to contend with the hostile forces of fact, indifference and romance disappear; that which he knows and can do, is exposed to severe tests; he can only use the hard and clear light of facts, and must give up the notion of lulling himself in agreeable illusions.

I rejoice, therefore, that I can once more address an assembly consisting almost exclusively of medical men who have gone through the same school. Medicine was once the intellectual home in which I grew up, and even the emigrant best understands and is best understood by his native land.

If I am called upon to designate in one word the fundamental error of that former time, I should be inclined to say that it pursued a false ideal of science in a one-sided and erroneous reverence for the deductive method. Medicine, it is true, was not the only science which was involved in this error, but in no other science have the consequences been so glaring, or have so hindered progress, as in medicine. The history of this science claims, therefore, a special interest in the history of the development of the human mind. None other is, perhaps, more fitted to show that a true criticism of the sources of cognition is also practically an exceedingly important object of true philosophy.

The proud word of Hippokrates,

ἰητρὸς φιλόσοφος ἰσόθεος,

'Godlike is the physician who is a philosopher,' served, as it were, as a banner of the old deductive medicine.

We may admit this if only we once agree what we

are to understand as a philosopher. For the ancients, philosophy embraced all theoretical knowledge; their philosophers pursued Mathematics, Physics, Astronomy, Natural History, in close connection with true philosophical or metaphysical considerations. If, therefore, we are to understand the medical philosopher of Hippokrates to be a man who has a *perfected* insight into the causal connection of natural processes, we shall in fact be able to say with Hippokrates, Such a one can give help like a god.

Understood in this sense, the aphorism describes in three words the ideal which our science has to strive after. But who can allege that it will ever attain this ideal?

But those disciples of medicine who thought themselves divine even in their own lifetime, and who wished to impose themselves upon others as such, were not inclined to postpone their hopes for so long a period. The requirements for the φιλόσοφος were considerably moderated. Every adherent of any given cosmological system, in which, for well or ill, facts must be made to correspond with reality, felt himself to be a philosopher. The philosophers of that time knew little more of the laws of Nature than the unlearned layman; but the stress of their endeavours was laid upon thinking, upon the logical consequence and completeness of the system. It is not difficult to understand how in periods of youthful development, such a one-sided over-estimate of thought could be arrived at. The superiority of man over animals, of the scholar over the barbarian, depends upon thinking; sensation, feeling, perception, on the contrary, he shares with his lower fellow-creatures, and in acuteness of the senses many of these are even superior to him. That man strives to develop his thinking faculty to the utmost is a problem on the solution of which the feeling of his own dignity, as well as of his own practical power, depends; and it is a natural error to have considered unimportant the dowry of mental capacities which Nature had given to animals, and to have believed that thought could be liberated from its natural basis, observation and perception, to begin its Icarian flight of metaphysical speculation.

It is, in fact, no easy problem to ascertain completely the origins of our knowledge. An enormous amount is transmitted by speech and writing. This power which man possesses of gathering together the stores of knowledge of generations, is the chief reason of his superiority over the animal, who is restricted to an inherited blind instinct and to its individual experience. But all transmitted knowledge is handed on already formed; whence the reporter has derived it, or how much criticism he has bestowed upon it, can seldom be made out, especially if the tradition has been handed down through several generations. We must admit it all upon good faith; we cannot arrive at the source; and when many generations have contented themselves with such knowledge, have brought no criticism to bear upon it; have, indeed, gradually added all kinds of small alterations, which ultimately grew up to large ones—after all this, strange things are often reported and believed under the authority of primeval wisdom. A curious case of this kind is the history of the circulation of the blood, of which we shall still have to speak.

But another kind of tradition by speech, which long remained undetected, is even still more confusing for one who reflects upon the origin of knowledge. Speech cannot readily develop names for classes of objects or for classes of processes, if we have not been accustomed very often to mention together the corresponding individuals, things, and separate cases, and to assert what there is in common about them. They must, therefore, possess many points in common. Or if we, reflecting scientifically upon this, select some of these characteristics, and collate them to form a definition, the common possession of these selected characteristics must necessitate that in the given cases a great number of other characteristics are to be regularly met with; there must be a natural connection between the first and the last-named characteristics. If, for instance, we assign the name of mammals to those animals which, when young, are suckled by their mothers, we can assert further, in reference to them, that they are all warm-blooded animals, born alive, that they have a spinal column but no quadrate bone, breathe through lungs, have separate divisions of the heart, &c. Hence the fact, that in the speech of an intelligent observing people a certain class of things are included in one name, indicates that these things or cases fall under a common natural relationship; by this alone a host of experiences are transmitted from preceding generations without this appearing to be the case.

The adult, moreover, when he begins to reflect upon the origin of his knowledge, is in possession of a huge mass of every-day experiences, which in great part reach back to the obscurity of his first childhood. Everything individual has long been forgotten, but the similar traces which the daily repetition of similar cases has left in his memory have deeply engraved themselves. And since only that which is in conformity with law is always repeated with regularity, these deeply impressed remains of all previous conceptions are just the conceptions of what is conformable to law in the things and processes.

Thus man, when he begins to reflect, finds that he possesses a wide range of acquirements of which he knows not whence they came, which he has possessed as long as he can remember. We need not refer even to the possibility of inheritance by procreation.

The conceptions which he has formed, which his mother tongue has transmitted, assert themselves as regulative powers, even in the objective world of fact, and as he does not know that he or his forefathers have developed these conceptions from the things themselves, the world of facts seems to him, like his conceptions, to be governed by intellectual forces. We recognise this psychological anthropomorphism, from the *Ideas* of Plato, to the immanent dialectic of the cosmical process of Hegel, and to the unconscious will of Schopenhauer.

Natural science, which in former times was virtually identical with medicine, followed the path of philosophy; the deductive method seemed to be capable of doing everything. Socrates, it is true, had developed the inductive conception in the most instructive manner. But the best which he accomplished remained virtually misunderstood.

I will not lead you through the motley confusion of pathological theories which, according to the varying inclination of their authors, sprouted up in consequence of this or the other increase of natural knowledge, and were mostly put forth by physicians, who obtained fame and renown as great observers and empirics, independently of their theories. Then came the less gifted pupils, who copied their master, exaggerated his theory, made it more one-sided and more logical, without regard to any discordance with Nature. The more rigid the system, the fewer and the more thorough were the methods to which the healing art was restricted. The more the schools were driven into a corner by the increase in actual knowledge, the more did they depend upon the ancient authorities, and the more intolerant were they against innovation. The great reformer of anatomy, Vesalius, was cited before the Theological faculty of Salamanca; Servetus was burned at Geneva along with his book, in which he described the circulation of the lungs; and the Paris faculty prohibited the teaching of Harvey's doctrine of the circulation of the blood in its lecture rooms.

At the same time the bases of the systems from which these schools started were mostly views on natural science which it would have been quite right to utilise within a narrow circle. What was not right was the delusion that it was more scientific to refer all diseases to one kind of explanation, than to several. What was called the solidar pathology wanted to deduce everything from the altered mechanism of the solid parts, especially from their altered tension; from the *strictum* and *laxum*, from tone and want of tone, and afterwards from strained or relaxed nerves and from obstructions in the vessels. Humoral pathology was only acquainted with alterations in mixture. The four cardinal fluids, representatives of the classical four elements, blood, phlegm, black and yellow gall; with others, the acrimonies or dyscrasies, which had to be expelled by sweating and purging; in the beginning of our modern epoch, the acids and alkalies or the alchymistic spirits, and the occult qualities of the substances assimilated—all these were the elements of this chemistry. Along with these were found all kinds of physiological conceptions, some of which contained remarkable foreshadowings, such as the ἔμφυτον θέρμον, the inherent vital force of Hippokrates, which is kept up by nutritive substances, this again boils in the stomach and is the source of all motion; here the thread is begun to be spun which subsequently led a physician to the law of the conservation of force. On the other hand, the πνεῦμα, which is half spirit and half air, which can be driven from the lungs into the arteries and fills them, has produced much confusion. The fact that air is generally found in the arteries of dead bodies, which indeed only penetrates in the moment in which the vessels are cut, led the ancients to the belief that air is also present in the arteries during life. The veins only remained then in which blood could circulate. It was believed to be formed in the liver, to move from there to the heart, and through the veins to the organs. Any careful observation of the operation of blood-letting must have taught that, in the veins, it comes from the periphery, and flows towards the heart. But this false theory had become so mixed up with the explanation of fever and of inflammation, that it acquired the authority of a dogma, which it was dangerous to attack.

Yet the essential and fundamental error of this system was, and still continued to be, the false kind of logical conclusion to which it was supposed to lead; the conception that it must be possible to build a complete system which would embrace all forms of disease, and their cure, upon any one such simple explanation. Complete knowledge of the causal connection of one class of phenomena gives finally a logical coherent system. There is no prouder edifice of the most exact thought than modern astronomy, deduced even to the minutest of its small disturbances, from Newton's law of gravitation. But Newton had been preceded by Kepler, who had by induction collated all the facts; and the astronomers have never believed that Newton's force excluded the simultaneous action of other forces. They have been continually on the watch to see whether friction, resisting media, and swarms of meteors have not also some influence. The older philosophers and physicians believed they could deduce, before they had settled their general principles by induction. They forgot that a deduction can have no more certainty than the principle from which it is deduced; and that each new induction must in the first place be a new test, by experience, of its own bases. That a conclusion is deduced by the strictest logical method from an uncertain premiss does not give it a hair's breadth of certainty or of value.

One characteristic of the schools which built up their system on such hypotheses, which they assumed as dogmas, is the intolerance of expression which I have already partially mentioned. One who works upon a well-ascertained foundation may readily admit an error; he loses, by so doing, nothing more than that in which he erred. If, however, the starting-point has been placed upon a hypothesis, which either appears guaranteed by authority, or is only chosen because it agrees with that which it is *wished* to believe true, any crack may then hopelessly destroy the whole fabric of conviction. The convinced disciples must therefore claim for each individual part of such a fabric the same degree of infallibility; for the anatomy of Hippokrates just as much as for fever crises; every opponent must only appear then as stupid or depraved, and the dispute will thus, according to old precedent, be so much the more passionate and personal, the more uncertain is the basis which is defended. We have frequent opportunities of confirming these general rules in the schools of dogmatic deductive medicine. They turned their intolerance partly against each other, and partly against the eclectics who found various explanations for various forms of disease. This method, which in its essence is completely justified, had, in the eyes of systematists, the defect of being illogical. And yet the greatest physicians and observers, Hippokrates at the head, Aretæus, Galen, Sydenham, and Boerhaave, had become eclectics, or at any rate very lax systematists.

About the time when we seniors commenced the study of medicine, it was still under the influence of the important discoveries which Albrecht von Haller had made on the excitability of nerves; and which he had placed in connection with the vitalistic theory of the nature of life. Haller had observed the excitability in the nerves and muscles of amputated members. The most surprising thing to him was, that the most varied external actions, mechanical, chemical, thermal, to which electrical ones were subsequently added, had always the same result; namely, that they produced muscular contraction. They were only quantitatively distinguished as regards their action on the organism, that is, only by the strength of the excitation; he designated them by the common name of *stimulus*; he called the altered condition of the nerve the *excitation*, and its capacity of responding to a stimulus the *excitability*, which was lost at death. This entire condition of things, which physically speaking asserts no more than that the nerves, as concerns the changes which take place in them after excitation, are in an exceedingly unstable state of equilibrium; this was looked upon as the fundamental property of animal life, and was unhesitatingly transferred to the other organs and tissues of the body, for which there was no similar justification. It was believed that none of them were active of themselves, but must receive an impulse by a stimulus from without; air and nourishment were considered to be the normal stimuli. The *kind* of activity seemed, on the contrary, to be conditioned by the specific energy of the organ, under the influence of the vital force. Increase or diminution of the excitability was the category under which the whole of the acute diseases were referred, and from which indications were taken as to whether the treatment should be lowering or stimulating. The rigid one-sidedness and the unrelenting logic with which Robert Brown had once worked out this system was broken, but it always furnished the leading points of view.

The vital force had formerly lodged as ethereal spirit, as a Pneuma in the arteries; it had then with Paracelsus acquired the form of an Archeus, a kind of useful Kobold, or indwelling alchymist, and had acquired its clearest scientific position as 'soul of life,' *anima inscia*, in Georg Ernst Stahl, who, in the first half of the last century, was professor of chemistry and pathology in Halle. Stahl had a clear and acute mind, which is informing and stimulating, from the way in which he states the proper question, even in those cases in which he decides against our present views. He it is who established the first comprehensive system of chemistry, that of phlogiston. If we translate his phlogiston into latent heat, the theoretical bases of his system passed essentially into the system of Lavoisier; Stahl did not then know oxygen, which occasioned some false hypotheses; for instance, on the negative gravity of phlogiston. Stahl's 'soul of life' is, on the whole, constructed on the pattern on which the pietistic communities of that period represented to themselves the sinful human soul; it is subject to errors and passions, to sloth, fear, impatience, sorrow, indiscretion, despair. The physician must first appease it, or then incite it, or punish it, and compel it to repent. And the way in which, at the same time, he established the necessity of the physical and vital actions was well thought out. The soul of life governs the body, and only acts by means of the physico-chemical forces of the substances assimilated. But it has the power to bind and to loose these forces, to allow them full play or to restrain them. After death the restrained forces become free, and evoke putrefaction and decomposition. For the refutation of this hypothesis of binding and loosing, it was necessary to discover the law of the conservation of force.

The second half of the previous century was too much possessed by the principles of rationalism to recognise openly Stahl's 'soul of life.' It was presented more scientifically as vital force, *Vis vitalis*, while in the main it retained its functions, and under the name of ' Nature's healing power ' it played a prominent part in the treatment of diseases.

The doctrine of vital force entered into the pathological system of changes in irritability. The attempt was made to separate the direct actions of the virus which produce disease, in so far as they depended on the play of blind natural forces, the *symptomata morbi*, from those which brought on the reaction of vital force, the *symptomata reactionis*. The latter were principally seen in inflammation and in fever. It was the function of the physician to observe the strength of this reaction, and to stimulate or moderate it according to circumstances.

The treatment of fever seemed at that time to be the chief point ; to be that part of medicine which had a real scientific foundation, and in which the local treatment fell comparatively into the background. The therapeutics of febrile diseases had thereby become very monotonous, although the means indicated by theory were still abundantly used, and especially blood-letting.

which since that time has almost been entirely abandoned. Therapeutics became still more impoverished as the younger and more critical generation grew up, and tested the assumptions of that which was considered to be scientific. Among the younger generation were many who, in despair as to their science, had almost entirely given up therapeutics, or on principle had grasped at an empiricism such as Rademacher then taught, which regarded any expectation of a scientific explanation as a vain hope.

What we learned at that time were only the ruins of the older dogmatism, but their doubtful features soon manifested themselves.

The vitalistic physician considered that the essential part of the vital processes did not depend upon natural forces, which, doing their work with blind necessity and according to a fixed law, determined the result. What these forces could do appeared quite subordinate, and scarcely worthy of a minute study. He thought that he had to deal with a soul-like being, to which a thinker, a philosopher, and an intelligent man must be opposed. May I elucidate this by a few outlines?

At this time auscultation and percussion of the organs of the chest were being regularly practised in the clinical wards. But I have often heard it maintained that they were a coarse mechanical means of investigation which a physician with a clear mental vision did not need; and it indeed lowered and debased the patient, who was anyhow a human being, by treating him as a machine. To feel the pulse seemed the most direct method of learning the mode of action of the vital force, and it was practised, therefore, as by far the most important means of investigation. To count with a repeater was quite usual, but seemed to the old gentlemen as a method not quite in good taste. There was, as yet, no idea of measuring temperature in cases of disease. In reference to the ophthalmoscope, a celebrated surgical colleague said to me that he would never use the instrument, it was too dangerous to admit crude light into diseased eyes; another said the mirror might be useful for physicians with bad eyes, his, however, were good, and he did not need it.

A professor of physiology of that time, celebrated for his literary activity, and noted as an orator and intelligent man, had a dispute on the images in the eye with his colleague the physicist. The latter challenged the physiologist to visit him and witness the experiment. The physiologist, however, refused his request with indignation; alleging that a physiologist had nothing to do with experiments; they were of no good but for the physicist. Another aged and learned professor of therapeutics, who occupied himself much with the reorganisation of the Universities, was urgent with me to divide physiology, in order to restore the good old time; that I myself should lecture on the really intellectual part, and should hand over the lower experimental part to a colleague whom he regarded as good enough for the purpose. He quite gave me up when I said that I myself considered experiments to be the true basis of science.

I mention these points, which I myself have experienced, to elucidate the feeling of the older schools, and indeed of the most illustrious representatives of medical science, in reference to the progressive set of ideas of the natural sciences; in literature these ideas naturally found feebler expression, for the old gentlemen were cautious and worldly wise.

You will understand how great a hindrance to progress such a feeling on the part of influential and respected men must have been. The medical education of that time was based mainly on the study of books; there were still lectures, which were restricted to mere dictation; for experiments and demonstrations in the laboratory the provision made was sometimes good and sometimes the reverse; there were no physiological and physical laboratories in which the student himself might go to work. Liebig's great deed, the foundation of the chemical laboratory, was complete, as far as chemistry was concerned, but his example had not been imitated elsewhere. Yet medicine possessed in anatomical dissections a great means of education for independent observation, which is wanting in the other faculties, and to which I am disposed to attach great weight. Microscopic demonstrations were isolated and infrequent in the lectures. Microscopic instruments were costly and scarce. I came into possession of one by having spent my autumn vacation in 1841 in the Charité, prostrated by typhoid fever; as pupil, I was nursed without expense, and on my recovery I found myself in possession of the savings of my small resources. The instrument was not beautiful, yet I was able to recognise by its means the prolongations of the ganglionic cells in the invertebrata, which I described in my dissertation, and to investigate the vibrions in my research on putrefaction and fermentation.

Any of my fellow-students who wished to make experiments had to do so at the cost of his pocket-money. One thing we learned thereby, which the younger generation does not, perhaps, learn so well in the laboratories—that is, to consider in all directions the ways and means of attaining the end, and to exhaust all possibilities, in the consideration, until a practicable path was found. We had, it is true, an almost uncultivated field before us, in which almost every stroke of the spade might produce remunerative results.

It was one man more especially who aroused our

enthusiasm for work in the right direction—that is, Johannes Müller, the physiologist. In his theoretical views he favoured the vitalistic hypothesis, but in the most essential points he was a natural philosopher, firm and immovable; for him, all theories were but hypotheses, which had to be tested by facts, and about which facts could alone decide. Even the views upon those points which most easily crystallise into dogmas, on the mode of activity of the vital force and the activity of the conscious soul, he tried continually to define more precisely, to prove or to refute by means of facts.

And, although the art of anatomical investigation was most familiar to him, and he therefore recurred most willingly to this, yet he worked himself into the chemical and physical methods which were more foreign to him. He furnished the proof that fibrine is dissolved in blood; he experimented on the propagation of sound in such mechanisms as are found in the drum of the ear; he treated the action of the eye as an optician. His most important performance for the physiology of the nervous system, as well as for the theory of cognition, was the actual definite establishment of the doctrine of the specific energies of the nerves. In reference to the separation of the nerves of motor and sensible energy, he showed how to make the experimental proof of Bell's law of the roots of the spinal cord so as to be free from errors; and in regard to the sensible energies he not only established the general law, but carried out a great number of separate investigations, to eliminate objections, and to refute false indications and evasions. That which hitherto had been imagined from the data of every-day experience, and which had been sought to be expressed in a vague manner, in which the true was mixed up with the false; or which had just been established for individual branches, such as by Dr. Young for the theory of colours, or by Sir Charles Bell for the motor nerves, that emerged from Müller's hands in a state of classical perfection—a scientific achievement whose value I am inclined to consider as equal to that of the discovery of the law of gravitation.

His scientific tendency, and more especially his example, were continued in his pupils. We had been preceded by Schwann, Henle, Reichert, Peters, Remak; I met as fellow-students E. Du Bois-Reymond, Virchow, Brücke, Ludwig, Traube, J. Meyer, Lieberkühn, Hallmann; we were succeeded by A. von Graefe, W. Busch, Max Schultze, A. Schneider.

Microscopic and pathological anatomy, the study of organic types, physiology, experimental pathology and therapeutics, ophthalmology, developed themselves in Germany under the influence of this powerful impulse far beyond the standard of rival adjacent countries. This was helped by the labours of those of similar tendencies among Müller's contemporaries, among whom the three brothers Weber of Leipzig must first of all be mentioned, who have built solid foundations in the mechanism of the circulation, of the muscles, of the joints, and of the ear.

The attack was made wherever a way could be perceived of understanding one of the vital processes; it was assumed that they could be understood, and success justified this assumption. A delicate and copious technical apparatus has been developed in the methods of microscopy, of physiological chemistry, and of vivisection; the latter greatly facilitated more particularly by the use of anæsthetic ether and of the paralysing curara, by which a number of deep problems became open to attack, which to our generation seemed hopeless. The thermometer, the ophthalmoscope, the auricular speculum, the laryngoscope, nervous irritation on the living body, opened out to the physician possibilities of delicate and yet certain diagnosis where there seemed to be absolute darkness. The continually increasing number of proved parasitical organisms substitute tangible objects for mystical entities, and teach the surgeon to forestall the fearfully subtle diseases of decomposition.

But do not think, gentlemen, that the struggle is at an end. As long as there are people of such astounding conceit as to imagine that they can effect, by a few clever strokes, that which man can otherwise only hope to achieve by toilsome labour, hypotheses will be started which, propounded as dogmas, at once promise to solve all riddles. And as long as there are people who believe implicitly in that which they wish to be true, so long will the hypotheses of the former find credence. Both classes will certainly not die out, and to the latter the majority will always belong.

There are two characteristics more particularly which metaphysical systems have always possessed. In the first place man is always desirous of feeling himself to be a being of a higher order, far beyond the standard of the rest of nature; this wish is satisfied by the spiritualists. On the other hand, he would like to believe that by his thought he was unrestrained lord of the world, and of course by his thinking with those conceptions, to the development of which he has attained; this is attempted to be satisfied by the materialists.

But one who, like the physician, has actively to face natural forces which bring about weal or woe, is also under the obligation of seeking for a knowledge of the truth, and of the truth only; without considering whether, what he finds, is pleasant in one way or the other. His aim is one which is firmly settled; for him the success of facts is alone finally decisive. He must endeavour to ascertain beforehand, what will be the result of his attack if he pursues this or that course.

In order to acquire this foreknowledge of what is coming, but of what has not been settled by observations, no other method is possible than that of endeavouring to arrive at the laws of facts by observations; and we can only learn them by induction, by the careful selection, collation, and observation of those cases which fall under the law. When we fancy that we have arrived at a law, the business of deduction commences. It is then our duty to develop the consequences of our law as completely as may be, but in the first place only to apply to them the test of experience, so far as they can be tested, and then to decide by this test whether the law holds, and to what extent. This is a test which really never ceases. The true natural philosopher reflects at each new phenomenon, whether the best established laws of the best known forces may not experience a change; it can of course only be a question of a change which does not contradict the whole store of our previously collected experiences. It never thus attains unconditional truth, but such a high degree of probability that it is practically equal to certainty. The metaphysicians may amuse themselves at this; we will take their mocking to heart when they are in a position to do better, or even as well. The old words of Socrates, the prime master of inductive definitions, in reference to them are just as fresh as they were 2,000 years ago : 'They imagined they knew what they did not know, and he at any rate had the advantage of not pretending to know what he did not know.' And again, he was surprised at its not being clear to them that it is not possible for men to discover such things; since even those who most prided themselves on the speeches made on the matter, did not agree among themselves, but behaved to each other like madmen (τοῖς μαινομένοις ὁμοίως).[1] Socrates calls them τοὺς μέγιστον φρονοῦντας. Schopenhauer[2] calls himself a Mont Blanc, by the side of a mole-heap, when he compares himself with a natural philosopher. The pupils admire these big words and try to imitate the master.

In speaking against the empty manufacture of hypotheses, do not by any means suppose that I wish to diminish the real value of original thoughts. The first discovery of a new law, is the discovery of a similarity which has hitherto been concealed in the course of natural processes. It is a manifestation of that which our forefathers in a serious sense described as 'wit'; it is of the same quality as the highest performances of artistic perception in the discovery of new types of expression. It is something which cannot be forced, and which cannot be acquired by any known method.

---

[1] Xenophon, *Memorabil.* I. i. 11.
[2] Arthur Schopenhauer, *Von ihm. über ihn von Fraucnstadt und Lindner.* Berlin, 1863, p. 653.

Hence all those aspire after it who wish to pass as the favoured children of genius. It seems, too, so easy, so free from trouble, to get by sudden mental flashes an unattainable advantage over our contemporaries. The true artist and the true inquirer knows that great works can only be produced by hard work. The proof that the ideas formed do not merely scrape together superficial resemblances, but are produced by a quick glance into the connection of the whole, can only be acquired when these ideas are completely developed—that is, for a newly discovered natural law, only by its agreement with facts. This estimate must by no means be regarded as depending on external success, but the success is here closely connected with the depth and completeness of the preliminary perceptions.

To find superficial resemblances is easy; it is amusing in society, and witty thoughts soon procure for their author the name of a clever man. Among the great number of such ideas, there must be some which are ultimately found to be partially or wholly correct; it would be a stroke of skill *always* to guess falsely. In such a happy chance a man can loudly claim his priority for the discovery; if otherwise, a lucky oblivion conceals the false conclusions. The adherents of such a process are glad to certify the value of a first thought. Conscientious workers who are shy at bringing their thoughts before the public before they have tested them in all directions, solved all doubts, and have firmly established the proof, these are at a decided disadvantage. To settle the present kind of questions of priority, only by the date of their first publication, and without considering the ripeness of the research, has seriously favoured this mischief.

In the 'type case' of the printer all the wisdom of the world is contained which has been or can be discovered; it is only requisite to know how the letters are to be arranged. So also, in the hundreds of books and pamphlets which are every year published about ether, the structure of atoms, the theory of perception, as well as on the nature of the asthenic fever and carcinoma, all the most refined shades of possible hypotheses are exhausted, and among these there must necessarily be many fragments of the correct theory. But who knows how to find them?

I insist upon this in order to make clear to you that all this literature, of untried and unconfirmed hypotheses, has no value in the progress of science. On the contrary, the few sound ideas which they may contain are concealed by the rubbish of the rest; and one who wants to publish something really new—facts—sees himself open to the danger of countless claims of priority, unless he is prepared to waste time and power in reading beforehand a quantity of absolutely useless books, and to destroy his readers' patience by a multitude of useless quotations.

Our generation has had to suffer under the tyranny of spiritualistic metaphysics; the newer generation will probably have to guard against that of the materialistic hypotheses. Kant's rejection of the claims of pure thought has gradually made some impression, but Kant allowed one way of escape. It was as clear to him as to Socrates that all metaphysical systems which up to that time had been propounded were tissues of false conclusions. His *Kritik der reinen Vernunft* is a continual sermon against the use of the category of thought beyond the limits of possible experience. But geometry seemed to him to do something which metaphysics was striving after; and hence geometrical axioms, which he looked upon as *à priori* principles antecedent to all experience, he held to be given by transcendental intuition, or as the inherent form of all external intuition. Since that time, pure *à priori* intuition has been the anchoring-ground of metaphysicians. It is even more convenient than pure thought, because everything can be heaped on it without going into chains of reasoning, which might be capable of proof or of refutation. The nativistic theory of perception of the senses is the expression of this theory in physiology. All mathematicians united to fight against any attempt to resolve the intuitions into their natural elements; whether the so-called pure or the empirical, the axioms of geometry, the principles of mechanics, or the perceptions of vision. For this reason, therefore, the mathematical investigations of Lobatschewsky, Gauss, and Riemann on the alterations which are logically possible in the axioms of geometry; and the proof that the axioms are principles which are to be confirmed or perhaps even refuted by experience, and can accordingly be acquired from experience—these I consider to be very important steps. That all metaphysical sects get into a rage about this must not lead you astray, for these investigations lay the axe at the bases of apparently the firmest supports which their claims still possess. Against those investigators who endeavour to eliminate from among the perceptions of the senses, whatever there may be of the actions of memory, and of the repetition of similar impressions, which occur in memory; whatever, in short, is a matter of experience, against them it is attempted to raise a party cry that they are spiritualists. As if memory, experience, and custom were not also facts, whose laws are to be sought, and which are not to be explained away because they cannot be glibly referred to reflex actions, and to the complex of the prolongation of ganglionic cells, and of the connection of nerve-fibres in the brain.

Indeed, however self-evident, and however important

the principle may appear to be, that natural science has to seek for the laws of facts, this principle is nevertheless often forgotten. In recognising the law found, as a force which rules the processes in nature, we conceive it objectively as a *force*, and such a reference of individual cases to a force which under given conditions produces a definite result, that we designate as a causal explanation of phenomena. We cannot always refer to the forces of atoms; we speak of a refractive force, of electromotive and of electrodynamic force. But do not forget the *given conditions* and the *given result*. If these cannot be given, the explanation attempted is merely a modest confession of ignorance, and then it is decidedly better to confess this openly.

If any process in vegetation is referred to forces in the cells, without a closer definition of the conditions among which, and of the direction in which, they work, this can at most assert that the more remote parts of the organism are without influence; but it would be difficult to confirm this with certainty in more than a few cases. In like manner, the originally definite sense which Johannes Müller gave to the idea of reflex action, is gradually evaporated into this, that when an impression has been made on any part of the nervous system, and an action occurs in any other part, this is supposed to have been explained by saying that it is a reflex action. Much may be imposed upon the irresolvable complexity of the nerve-fibres of the brain But the resemblance to the *qualitates occultæ* of ancient medicine is very suspicious.

From the entire chain of my argument it follows that what I have said against metaphysics is not intended against philosophy. But metaphysicians have always tried to plume themselves on being philosophers, and philosophical amateurs have mostly taken an interest in the high-flying speculations of the metaphysicians, by which they hope in a short time, and at no great trouble, to learn the whole of what is worth knowing. On another occasion [1] I compared the relationship of metaphysics to philosophy with that of astrology to astronomy. The former had the most exciting interest for the public at large, and especially for the fashionable world, and turned its alleged connoisseurs into influential persons. Astronomy, on the contrary, although it had become the ideal of scientific research, had to be content with a small number of quietly working disciples.

In like manner, philosophy, if it gives up metaphysics, still possesses a wide and important field, the knowledge of mental and spiritual processes and their laws. Just as the anatomist, when he has reached the limits of microscopic vision, must try to gain an in-

[1] Preface to the German translation of Tyndall's *Scientific Fragments*, p. xxii.

sight into the action of his optical instrument, in like manner every scientific enquirer must study minutely the chief instrument of his research as to its capabilities. The groping of the medical schools for the last two thousand years is, among other things, an illustration of the harm of erroneous views in this respect. And the physician, the statesman, the jurist, the clergyman, and the teacher, ought to be able to build upon a knowledge of physical processes if they wish to acquire a true scientific basis for their practical activity. But the true science of philosophy has had, perhaps, to suffer more from the evil mental habits and the false ideals of metaphysics than even medicine itself.

One word of warning. I should not like you to think that my statements are influenced by personal irritation. I need not explain that one who has such opinions as I have laid before you, who impresses on his pupils, whenever he can, the principle that 'a metaphysical conclusion is either a false conclusion or a concealed experimental conclusion,' that he is not exactly beloved by the votaries of metaphysics or of intuitive conceptions. Metaphysicians, like all those who cannot give any decisive reasons to their opponents, are usually not very polite in their controversy; one's own success may approximately be estimated from the increasing want of politeness in the replies.

My own researches have led me more than other disciples of the school of natural science into controversial regions; and the expressions of metaphysical discontent have perhaps concerned me even more than my friends, as many of you are doubtless aware.

In order, therefore, to leave my own personal opinions quite on one side, I have allowed two unsuspected warrantors to speak for me—Socrates and Kant—both of whom were certain that all metaphysical systems established up to their time were full of empty false conclusions, and who guarded themselves against adding any new ones. In order to show that the matter has not changed, either in the last 2,000 years or in the last 100 years, let me conclude with a sentence of one who was unfortunately too soon taken away from us, Frederick Albert Lange, the author of the 'History of Materialism.' In his posthumous 'Logical Studies,' which he wrote in anticipation of his approaching end, he gives the following picture, which struck me because it would hold just as well in reference to solidar or humoral pathologists, or any other of the old dogmatic schools of medicine.

Lange says: The Hegelian ascribes to the Herbartian a less perfect knowledge than to himself, and conversely; but neither hesitates to consider the knowledge of the other to be higher compared with that of the empiricist, and to recognise in it at any rate an approximation to the only true knowledge. It is seen, also, that here no regard is paid to the validity of the proof, and that a mere statement in the form of a deduction from the entirety of a system is recognised as 'apodictic knowledge.'

Let us, then, throw no stones at our old medical predecessors, who in dark ages, and with but slight preliminary knowledge, fell into precisely the same errors as the great intelligences of what wishes to be thought the illuminated nineteenth century. They did no worse than their predecessors except that the nonsense of their method was more prominent in the matter of natural science. Let us work on. In this work of true intelligence physicians are called upon to play a prominent part. Among those who are continually called upon actively to preserve and apply their knowledge of nature, you are those who begin with the best mental preparation, and are acquainted with the most varied regions of natural phenomena.

In order, finally, to conclude our consultation on the condition of Dame Medicine correctly with the epikrisis, I think we have every reason to be content with the success of the treatment which the school of natural science has applied, and we can only recommend the younger generation to continue the same therapeutics.

ON ACADEMIC FREEDOM IN GERMAN UNIVERSITIES.

Inaugural Address as Rector of the Frederick William University of Berlin. Delivered October 15, 1877.

In entering upon the honourable office to which the confidence of my colleagues has called me, my first duty is once more openly to express my thanks to those who have thus honoured me by their confidence. I have the more reason to appreciate it highly, as it was conferred upon me, notwithstanding that I have been but few years among you, and notwithstanding that I belong to a branch of natural science which has come within the circle of University instruction in some sense as a foreign element; which has necessitated many changes in the old order of University teaching, and which will, perhaps, necessitate other changes. It is indeed just in that branch (Physics) which I represent, and which forms the theoretical basis of all other branches of Natural Science, that the particular characteristics of their methods are most definitely pronounced. I have already been several times in the position of having to propose alterations in the previous regulations of the University, and I have always had the pleasure of meeting with the ready assistance of my colleagues in the faculty, and of the Senate. That you have made me the Director of the business of this University for this year, is a proof that you regard me as no thoughtless innovator. For, in fact, however the objects, the methods, the more immediate aims of investigations in the natural sciences may differ externally from those of the mental sciences, and however foreign their results and however remote their interest may often appear, to those who are accustomed only to the direct manifestations and products of mental activity, there is in reality, as I have endeavoured to show in my discourse as Rector at Heidelberg, the closest connection in the essentials of scientific methods, as well as in the ultimate aims of both classes of the sciences. Even if most of the objects of investigation of the natural sciences are not directly connected with the interests of the mind, it cannot, on the other hand, be forgotten that the power of true scientific method stands out in the natural sciences far more prominently—that the real is far more sharply separated from the unreal, by the incorruptible criticism of facts, than is the case with the more complex problems of mental science.

And not merely the development of this new side of scientific activity, which was almost unknown to antiquity, but also the influence of many political, social, and even international relationships make themselves felt, and require to be taken into account. The circle of our students has had to be increased; a changed national life makes other demands upon those who are leaving; the sciences become more and more specialised and divided; exclusive of the libraries, larger and more varied appliances for study are required. We can scarcely foresee what fresh demands and what new problems we may have to meet in the more immediate future.

On the other hand, the German Universities have conquered a position of honour not confined to their fatherland; the eyes of the civilised world are upon them. Scholars speaking the most different languages crowd towards them, even from the farthest parts of the earth. Such a position would be easily lost by a false step, but would be difficult to regain.

Under these circumstances it is our duty to get a clear understanding of the reason for the previous prosperity of our Universities; we must try to find what is the feature in their arrangements which we must seek to retain as a precious jewel, and where, on the contrary, we may give way when changes are required. I consider myself by no means entitled to give a final opinion on this matter. The point of view of any single individual is restricted; representatives of other sciences will be able to contribute something. But I think that a final result can only be arrived at when each one becomes clear as to the state of things as seen from his point of view.

The European Universities of the Middle Age had their origin as free private unions of their students, who came together under the influence of celebrated teachers, and themselves arranged their own affairs. In recognition of the public advantage of these unions they soon obtained from the State, privileges and honourable rights, especially that of an independent jurisdiction, and the right of granting academic degrees. The students of that time were mostly men of mature years, who frequented the University more immediately for their own instruction and without any direct practical object; but younger men soon began to be sent, who, for the most part, were placed under the superintendence of the older members. The separate Universities split again into closer economic unions, under the name of 'Nations,' 'Bursaries,' 'Colleges,' whose older members, the seniors, governed the common affairs of each such union, and also met together for regulating the common affairs of the University. In the courtyard of the University of Bologna are still to be seen the coats-of-arms, and lists of members and seniors, of many such Nations in ancient times. The older graduated members were regarded as permanent life members of such Unions, and they retained the right of voting, as is still the case in the College of Doctors in the University of Vienna, and in the Colleges of Oxford and of Cambridge, or was until recently.

Such a free confederation of independent men, in which teachers as well as taught were brought together by no other interest than that of love of science; some by the desire of discovering the treasure of mental culture which antiquity had bequeathed, others endeavouring to kindle in a new generation the ideal enthusiasm which had animated their lives. Such was the origin of Universities, based, in the conception, and in the plan of their organisation, upon the most perfect freedom. But we must not think here of freedom of teaching in the modern sense. The majority was usually very intolerant of divergent opinions. Not unfrequently the adherents of the minority were compelled to quit the University in a body. This was not restricted to those cases in which the Church intermeddled, and where political or metaphysical propositions were in question. Even the medical faculties—

that of Paris, the most celebrated of all at the head—allowed no divergence from that which they regarded as the teaching of Hippocrates. Anyone who used the medicines of the Arabians or who believed in the circulation of the blood was expelled.

The change, in the Universities, to their present constitution, was caused mainly by the fact that the State granted to them material help, but required, on the other hand, the right of co-operating in their management. The course of this development was different in different European countries, partly owing to divergent political conditions and partly to that of national character.

Until lately, it might have been said that the least change has taken place in the old English Universities, Oxford and Cambridge. Their great endowments, the political feeling of the English for the retention of existing rights, had excluded almost all change, even in directions in which such change was urgently required. Until of late both Universities had in great measure retained their character as schools for the clergy, formerly of the Roman and now of the Anglican Church, whose instruction laymen might also share in so far as it could serve the general education of the mind; they were subjected to such a control and mode of life, as was formerly considered to be good for young priests. They lived, as they still live, in colleges, under the superintendence of a number of older graduate members (Fellows) of the College; in other respects in the style and habits of the well-to-do classes in England.

The range and the method of the instruction is a more highly developed gymnasial instruction; though in its limitation to what is afterwards required in the examination, and in the minute study of the contents of prescribed text-books, it is more like the Repetitoria which are here and there held in our Universities. The acquirements of the students are controlled by searching examinations for academical degrees, in which very special knowledge is required, though only for limited regions. By such examinations the academical degrees are acquired.

While the English Universities give but little for the endowment of the positions of approved scientific teachers, and do not logically apply even that little for this object, they have another arrangement which is apparently of great promise for scientific study, but which has hitherto not effected much; that is the institution of Fellowships. Those who have passed the best examinations are elected as Fellows of their college, where they have a home, and along with this, a respectable income, so that they can devote the whole of their leisure to scientific pursuits. Both Oxford and Cambridge have each more than 500 such fellowships. The Fellows *may*, but *need* not act as tutors for the students. They need not even live in the University Town, but may spend their stipends where they like, and in many cases may retain the fellowships for an indefinite period. With some exceptions, they only lose it in case they marry, or are elected to certain offices. They are the real successors of the old corporation of students, by and for which the University was founded and endowed. But however beautiful this plan may seem, and notwithstanding the enormous sums devoted to it, in the opinion of all unprejudiced Englishmen it does but little for science; manifestly because most of these young men, although they are the pick of the students, and in the most favourable conditions possible for scientific work, have in their student-career not come sufficiently in contact with the living spirit of inquiry, to work on afterwards on their own account, and with their own enthusiasm.

In certain respects the English Universities do a great deal. They bring up their students as cultivated men, who are expected not to break through the restrictions of their political and ecclesiastical party, and, in fact, do not thus break through. In two respects we might well endeavour to imitate them. In the first place, together with a lively feeling for the beauty and youthful freshness of antiquity, they develop in a high degree a sense for delicacy and precision in writing which shows itself in the way in which they handle their mother-tongue. I fear that one of the weakest sides in the instruction of German youth is in this direction. In the second place the English Universities, like their schools, take greater care of the bodily health of their students. They live and work in airy, spacious buildings, surrounded by lawns and groves of trees; they find much of their pleasure in games which excite a passionate rivalry in the development of bodily energy and skill, and which in this respect are far more efficacious than our gymnastic and fencing exercises. It must not be forgotten that the more young men are cut off from fresh air and from the opportunity of vigorous exercise, the more induced will they be to seek an apparent refreshment in the misuse of tobacco and of intoxicating drinks. It must also be admitted that the English Universities accustom their students to energetic and accurate work, and keep them up to the habits of educated society. The *moral* effect of the more rigorous control is said to be rather illusory.

The Scotch Universities and some smaller English foundations of more recent origin—University College and King's College in London, and Owens College in Manchester—are constituted more on the German and Dutch model.

The development of French Universities has been quite different, and indeed almost in the opposite direction. In accordance with the tendency of the French to throw overboard everything of historic development to suit some rationalistic theory, their faculties have logically become purely institutes for instruction—special schools, with definite regulations for the course of instruction, developed and quite distinct from those institutions which are to further the progress of science, such as the *Collège de France*, the *Jardin des Plantes*, and the *École des Études Supérieures*. The faculties are entirely separated from one another, even when they are in the same town. The course of study is definitely prescribed, and is controlled by frequent examinations. French teaching is confined to that which is clearly established, and transmits this in a well-arranged, well worked-out manner, which is easily intelligible, and does not excite doubt nor the necessity for deeper inquiry. The teachers need only possess good receptive talents. Thus in France it is looked upon as a false step when a young man of promising talent takes a professorship in a faculty in the provinces. The method of instruction in France is well adapted to give pupils, of even moderate capacity, sufficient knowledge for the routine of their calling. They have no choice between different teachers, and they swear *in verba magistri*; this gives a happy self-satisfaction and freedom from doubts. If the teacher has been well chosen, this is sufficient in ordinary cases, in which the pupil does what he has seen his teacher do. It is only unusual cases that test how much actual insight and judgment the pupil has acquired. The French people are moreover gifted, vivacious, and ambitious, and this corrects many defects in their system of teaching.

A special feature in the organisation of French Universities consists in the fact that the position of the teacher is quite independent of the favour of his hearers; the pupils who belong to his faculty are generally compelled to attend his lectures, and the far from inconsiderable fees which they pay flow into the chest of the Minister of Education; the regular salaries of the University professors are defrayed from this source; the State gives but an insignificant contribution towards the maintenance of the University. When, therefore, the teacher has no real pleasure in teaching, or is not ambitious of having a number of pupils, he very soon becomes indifferent to the success of his teaching, and is inclined to take things easily.

Outside the lecture-rooms, the French students live without control, and associate with young men of other callings, without any special *esprit de corps* or common feeling.

The development of the German Universities differs characteristically from these two extremes. Too poor in their own possessions not to be compelled, with increasing demands for the means of instruction, eagerly to accept the help of the State, and too weak to resist encroachments upon their ancient rights in times in which modern States attempt to consolidate themselves, the German Universities have had to submit themselves to the controlling influence of the State. Owing to this latter circumstance the decision in all important University matters has in principle been transferred to the State, and in times of religious or political excitement this supreme power has occasionally been unscrupulously exerted. But in most cases the States which were working out their own independence were favourably disposed towards the Universities; they required intelligent officials, and the fame of their country's University conferred a certain lustre upon the Government. The ruling officials were, moreover, for the most part students of the University; they remained attached to it. It is very remarkable how among wars and political changes in the States fighting with the decaying Empire for the consolidation of their young sovereignties, while almost all other privileged orders were destroyed, the Universities of Germany saved a far greater nucleus of their internal freedom and of the most valuable side of this freedom, than in conscientious Conservative England, and than in France with its wild chase after freedom.

We have retained the old conception of students, as

that of young men responsible to themselves, striving after science of their own free will, and to whom it is left to arrange their own plan of studies as they think best. If attendance on particular lectures was enjoined for certain callings—what are called 'compulsory lectures'—these regulations were not made by the University, but by the State, which was afterwards to admit candidates to these callings. At the same time the students had, and still have, perfect freedom to migrate from one German University to another, from Dorpat to Zurich, from Vienna to Gratz; and in each University they had free choice among the teachers of the same subject, without reference to their position as ordinary or extraordinary professors or as private docents. The students are, in fact, free to acquire any part of their instruction from books; it is highly desirable that the works of great men of past times should form an essential part of study.

Outside the University there is no control over the proceedings of the students, so long as they do not come in collision with the guardians of public order. Beyond these cases the only control to which they are subject is that of their colleagues, which prevents them from doing anything which is repugnant to the feeling of honour of their own body. The Universities of the Middle Ages formed definite close corporations, with their own jurisdiction, which extended to the right over life and death of their own members. As they lived for the most part on foreign soil, it was necessary to have their own jurisdiction, partly to protect the members from the caprices of foreign judges, partly to keep up that degree of respect and order, within the society, which was necessary to secure the continuation of the rights of hospitality on a foreign soil; and partly, again, to settle disputes among the members. In modern times the remains of this academic jurisdiction have by degrees been completely transferred to the ordinary courts, or will be so transferred; but it is still necessary to maintain certain restrictions on a union of strong and spirited young men, which guarantee the peace of their fellow-students and that of the citizens. In cases of collision this is the object of the disciplinary power of the University authorities. This object, however, must be mainly attained by the sense of honour of the students; and it must be considered fortunate that German students have retained a vivid sense of corporate union, and of what is intimately connected therewith, a requirement of honourable behaviour in the individual. I am by no means prepared to defend every individual regulation in the Codex of Students' Honour; there are many Middle Age remains among them which were better swept away; but that can only be done by the students themselves.

For most foreigners the uncontrolled freedom of German students is a subject of astonishment; the more so as it is usually some obvious excrescences of this freedom which first meet their eyes; they are unable to understand how young men can be so left to themselves without the greatest detriment. The German looks back to his student life as to his golden age; our literature and our poetry are full of expressions of this feeling. Nothing of this kind is but even faintly suggested in the literature of other European peoples. The German student alone has this perfect joy in the time, in which, in the first delight in youthful responsibility, and freed more immediately from having to work for extraneous interests, he can devote himself to the task of striving after the best and noblest which the human race has hitherto been able to attain in knowledge and in speculation, closely joined in friendly rivalry with a large body of associates of similar aspirations, and in daily mental intercourse with teachers from whom he learns something of the workings of the thoughts of independent minds.

When I think of my own student life, and of the impression which a man like Johannes Müller, the physiologist, made upon us, I must place a very high value upon this latter point. Anyone who has once come in contact with one or more men of the first rank must have had his whole mental standard altered for the rest of his life. Such intercourse is, moreover, the most interesting that life can offer.

You, my younger friends, have received in this freedom of the German students a costly and valuable inheritance of preceding generations. Keep it—and hand it on to coming races, purified and ennobled if possible. You have to maintain it, by each, in his place, taking care that the body of German students is worthy of the confidence which has hitherto accorded such a measure of freedom. But freedom necessarily implies responsibility. It is as injurious a present for weak, as it is valuable for strong characters. Do not wonder if parents and statesmen sometimes urge that a more rigid system of supervision and control, like that of the English, shall be introduced even among us. There is no doubt that, by such a system, many a one would be saved who is ruined by freedom. But the State and the Nation is best served by those who can bear freedom, and have shown that they know how to work and to struggle, from their own force and insight and from their own interest in science.

My having previously dwelt on the influence of mental intercourse with distinguished men, leads me to discuss another point in which German Universities are distinguished from the English and French ones. It is that we start with the object of having instruction given, if possible, only by teachers who have proved their own power of advancing science. This also is a point in respect to which the English and French often express their surprise. They lay more weight than the Germans on what is called the 'talent for teaching'— that is, the power of explaining the subjects of instruction in a well-arranged and clear manner, and, if possible, with eloquence, and so as to entertain and to fix the attention. Lectures of eloquent orators at the Collège de France, Jardin des Plantes, as well as in Oxford and Cambridge, are often the centres of the elegant and the educated world. In Germany we are not only indifferent to, but even distrustful of, oratorical ornament, and often enough are more negligent than we should be of the outer forms of the lecture. There can be no doubt that a good lecture can be followed with far less exertion than a bad one; that the matter of the first can be more certainly and completely apprehended; that a well-arranged explanation, which develops the salient points and the divisions of the subject, and which brings it, as it were, almost intuitively before us, can impart more information in the same time than one which has the opposite qualities. I am by no means prepared to defend what is, frequently, our too great contempt for form in speech and in writing. It cannot also be doubted that many original men, who have done considerable scientific work, have often an uncouth, heavy, and hesitating delivery. Yet I have not infrequently seen that such teachers had crowded lecture-rooms, while empty-headed orators excited astonishment in the first lecture, fatigue in the second, and were deserted in the third. Anyone who desires to give his hearers a perfect conviction of the truth of his principles must, first of all, know from his own experience how conviction is acquired and how not. He must have known how to acquire conviction where no predecessor had been before him— that is, he must have worked at the confines of human knowledge, and have conquered for it new regions. A teacher who retails convictions which are foreign to him, is sufficient for those pupils who depend upon authority as the source of their knowledge, but not for such as require bases for their conviction which extend to the very bottom.

You will see that this is an honourable confidence which the nation reposes in you. Definite courses and specified teachers are not prescribed to you. You are regarded as men whose unfettered conviction is to be gained; who know how to distinguish what is essential from what is only apparent; who can no longer be appeased by an appeal to any authority, and who no longer let themselves be so appeased. Care is also always taken that you yourselves should penetrate to the sources of knowledge in so far as these consist in books and monuments, or in experiments, and in the observation of natural objects and processes.

Even the smaller German Universities have their own libraries, collections of casts, and the like. And in the establishment of laboratories for chemistry, microscopy, physiology, and physics, Germany has preceded all other European countries, who are now beginning to emulate her. In our own University we may in the next few weeks expect the opening of two new institutions devoted to instruction in natural science.

The free conviction of the student can only be acquired when freedom of expression is guaranteed to the teacher's own conviction—the *liberty of teaching*. This has not always been ensured, either in Germany or in the adjacent countries. In times of political and ecclesiastical struggle the ruling parties have often enough allowed themselves to encroach; this has always been regarded by the German nation as an attack upon their sanctuary. The advanced political freedom of the new German Empire has brought a cure for this. At this moment, the most extreme consequences of materialistic metaphysics, the boldest speculations upon the basis of Darwin's theory of evolution, may be taught in German Universities with as little restraint as the most extreme deification of Papal Infallibility. As in the tribune of European Parliaments it is forbidden to suspect motives or indulge in abuse of the personal qualities of our opponents, so also is any incitement to such acts as are legally forbidden. But there is no obstacle to the discussion of a scientific question in a scientific spirit. In English and French Universities there is less idea of liberty of teaching in this sense. Even in the Collège de France the lectures of a man of Renan's scientific importance and earnestness are forbidden.

I have to speak of another aspect of our liberty of teaching. That is, the extended sense in which German Universities have admitted teachers. In the original meaning of the word, a doctor is a 'teacher,' or one whose capacity as teacher is recognised. In the Universities of the Middle Ages any doctor who found pupils could set up as teacher. In course of time the practical signification of the title was changed. Most of those who sought the title did not intend to act as teachers, but only needed it as an official recognition of their scientific training. Only in Germany are there any remains of this ancient right. In accordance with the altered meaning of the title of doctor, and the minuter specialisation of the subjects of instruction, a special proof of more profound scientific proficiency, in the particular branch in which they wish to habilitate, is required from those doctors who desire to exercise the right of teaching. In most German

Universities, moreover, the legal status of these habilitated doctors as teachers is exactly the same as that of the ordinary professors. In a few places they are subject to some slight restrictions which, however, have scarcely any practical effect. The senior teachers of the University, especially the ordinary professors, have this amount of favour, that, on the one hand, in those branches in which special apparatus is needed for instruction, they can more freely dispose of the means belonging to the State; while on the other it falls to them to hold the examinations in the faculty, and, as a matter of fact, often also the State examination. This naturally exerts a certain pressure on the weaker minds among the students. The influence of examinations is, however, often exaggerated. In the frequent migrations of our students, a great number of examinations are held in which the candidates have never attended the lectures of the examiners.

On no feature of our University arrangements do foreigners express their astonishment so much as about the position of private docents. They are surprised, and even envious, that we have such a number of young men who, without salary, for the most part with insignificant incomes from fees, and with very uncertain prospects for the future, devote themselves to strenuous scientific work. And, judging us from the point of view of basely practical interests, they are equally surprised that the faculties so readily admit young men who at any moment may change from assistants to competitors; and further, that only in the most exceptional cases is anything ever heard of unworthy means of competition in what is a matter of some delicacy.

The appointment to vacant professorships, like the admission of private docents, rests, though not unconditionally, and not in the last resort, with the faculty, that is with the body of ordinary professors. These form, in German Universities, that residuum of former colleges of doctors to which the rights of the old corporations have been transferred. They form as it were a select committee of the graduates of a former epoch, but established with the co-operation of the Government. The usual form for the nomination of new ordinary professors is that the faculty proposes three candidates to Government for its choice; where the Government, however, does not consider itself restricted to the candidates proposed. Excepting in times of heated party conflict it is very unusual for the proposals of the faculty to be passed over. If there is not a very obvious reason for hesitation it is always a serious personal responsibility for the executive officials to elect, in opposition to the proposals of competent judges, a teacher who has publicly to prove his capacity before large circles.

The professors have, however, the strongest motives for securing to the faculty the best teachers. The most essential condition for being able to work with pleasure at the preparation of lectures is the consciousness of having not too small a number of intelligent listeners; moreover, a considerable fraction of the income of many teachers depends upon the number of their hearers. Each one must wish that his faculty, as a whole, shall attract as numerous and as intelligent a body of students as possible. That, however, can only be attained by choosing as many able teachers, whether professors or docents, as possible. On the other hand, a professor's attempt to stimulate his hearers to vigorous and independent research can only be successful when it is supported by his colleagues; besides this, working with distinguished colleagues makes life in University circles interesting, instructive, and stimulating. A faculty must have greatly sunk, it must not only have lost its sense of dignity, but also even the most ordinary worldly prudence, if other motives could preponderate over these; and such a faculty would soon ruin itself.

With regard to the spectre of rivalry among University teachers with which it is sometimes attempted to frighten public opinion, there can be none such if the students and their teachers are of the right kind. In the first place, it is only in large Universities that there are two to teach one and the same branch; and even if there is no difference in the official definition of the subject, there will be a difference in the scientific tendencies of the teachers; they will be able to divide the work in such a manner that each has that side which he most completely masters. Two distinguished teachers who are thus complementary to each other, form then so strong a centre of attraction for the students that both suffer no loss of hearers, though they may have to share among themselves a number of the less zealous ones.

The disagreeable effects of rivalry will be feared by a teacher who does not feel quite certain in his scientific position. This can have no considerable influence on the official decisions of the faculty when it is only a question of one, or of a small number, of the voters.

The predominance of a distinct scientific school in a faculty may become more injurious than such personal interests. When the school has scientifically outlived itself, students will probably migrate by degrees to other Universities. This may extend over a long period, and the faculty in question will suffer during that time.

We see best how strenuously the Universities under this system have sought to attract the scientific ability of Germany when we consider how many pioneers have remained outside the Universities. The answer to such an inquiry is given in the not infrequent jest or sneer that all wisdom in Germany is professorial wisdom. If we look at England, we see men like Humphry Davy, Faraday, Mill, Grote, who have had no connection with English Universities. If, on the other hand, we deduct from the list of German men of science those who, like David Strauss, have been driven away by Government for ecclesiastical or for political reasons, and those who, as members of learned Academies, had the right to deliver lectures in the Universities, as Alexander and Wilhelm von Humboldt, Leopold von Buch, and others, the rest will only form a small fraction of the number of the men of equal scientific standing who have been at work in the Universities; while the same calculation made for England would give exactly the opposite result. I have often wondered that the Royal Institution of London, a private Society, which provides for its members and others short courses of lectures on the Progress of Natural Science, should have been able to retain permanently the services of men of such scientific importance as Humphry Davy and Faraday. It was no question of great emoluments; these men were manifestly attracted by a select public consisting of men and women of independent mental culture. In Germany the Universities are unmistakably the institutions which exert the most powerful attraction on the taught. But it is clear that this attraction depends on the teacher's hope that he will not only find in the University a body of pupils enthusiastic and accustomed to work, but such also as devote themselves to the formation of an independent conviction. It is only with such students that the intelligence of the teacher bears any further fruit.

The entire organisation of our Universities is thus permeated by this respect for a free independent conviction, which is more strongly impressed on the Germans than on their Aryan kindred of the Celtic and Romanic branches, in whom practical political motives have greater weight. They are able, and as it would seem with perfect conscientiousness, to restrain the inquiring mind from the investigation of those principles which appear to them to be beyond the range of discussion, as forming the foundation of their political, social, and religious organisation; they think themselves quite justified in not allowing their youth to look beyond the boundary which they themselves are not disposed to overstep.

If, therefore, any region of questions is to be considered as outside the range of discussion, however remote and restricted it may be, and however good may be the intention, the pupils must be kept in the prescribed path, and teachers must be appointed who do not rebel against authority. We can then, however, only speak of free conviction in a very limited sense.

You see how different was the plan of our forefathers. However violently they may at times have interfered with individual results of scientific inquiry, they never wished to pull it up by the roots. An opinion which was not based upon independent conviction appeared to them of no value. In their hearts they never lost faith that freedom alone could cure the errors of freedom, and a riper knowledge the errors of what is unripe. The same spirit which overthrew the yoke of the Church of Rome, also organised the German Universities.

But any institution based upon freedom must also be able to calculate on the judgment and reasonableness of those to whom freedom is granted. Apart from the points which have been previously discussed, where the students themselves are left to decide on the course of their studies and to select their teachers, the above considerations show how the students react upon their teachers. To produce a good course of lectures is a labour which is renewed every term. New matter is continually being added which necessitates a reconsideration and a rearrangement of the old from fresh points of view. The teacher would soon be dispirited in his work if he could not count upon the zeal and the interest of his hearers. The estimate which he places on his task will depend on how far he is followed by the appreciation of a sufficient number of, at any rate, his more intelligent hearers. The influx of hearers to the lectures of a teacher has no slight influence upon his fame and promotion, and, therefore, upon the composition of the body of teachers. In all these respects, it is assumed that the general public opinion among the students cannot go permanently wrong. The majority of them—who are, as it were, the representatives of the general opinion—must come to us with a sufficiently logically trained judgment, with a sufficient habit of mental exertion, with a tact sufficiently developed on the best models, to be able to discriminate truth from the babbling appearance of truth. Among the students are to be found those intelligent heads who will be the mental leaders of the next generation, and who, perhaps, in a few years, will direct to themselves the eyes of the world. Occasional errors in youthful and excitable spirits naturally occur; but, on the whole, we may be pretty sure that they will soon set themselves right.

Thus prepared, they have hitherto been sent to us by the Gymnasiums. It would be very dangerous for the Universities if large numbers of students frequented them, who were less developed in the above respects. The general self-respect of the students must not be allowed to sink. If that were the case, the dangers of academic freedom would choke its blessings. It must therefore not be looked upon as pedantry, or arrogance, if the Universities are scrupulous in the admission of students of a different style of education. It would be still more dangerous if, for any extraneous reasons, teachers were introduced into the faculty, who have not the complete qualifications of an independent academical teacher.

Do not forget, my dear colleagues, that you are in a responsible position. You have to preserve the noble inheritance of which I have spoken, not only for your own people, but also as a model to the widest circles of humanity. You will show that youth also is enthusiastic, and will work for independence of conviction. I say work; for independence of conviction is not the facile assumption of untested hypotheses, but can only be acquired as the fruit of conscientious inquiry and strenuous labour. You must show that a conviction which you yourselves have worked out is a more fruitful germ of fresh insight, and a better guide for action, than the best-intentioned guidance by authority. Germany—which in the sixteenth century first revolted for the right of such conviction, and gave its witness in blood—is still in the van of this fight. To Germany has fallen an exalted historical task, and in it you are called upon to co-operate.

HERMANN VON HELMHOLTZ

AN

AUTOBIOGRAPHICAL SKETCH.

An Address delivered on the occasion of his Jubilee, 1891.

IN the course of the past year, and most recently on the occasion of the celebration of my seventieth birthday, and the subsequent festivities, I have been overloaded with honours, with marks of respect and of goodwill in a way which could never have been expected. My own sovereign, his Majesty the German Emperor, has raised me to the highest rank in the Civil Service; the Kings of Sweden and of Italy, my former sovereign, the Grand Duke of Baden, and the President of the French Republic, have conferred Grand Crosses on me; many academies, not only of science, but also of the fine arts, faculties, and learned societies spread over the whole world, from Tomsk to Melbourne, have sent me diplomas, and richly illuminated addresses, expressing in elevated language their recognition of my scientific endeavours, and their thanks for those endeavours, in terms which I cannot read without a feeling of shame. My native town, Potsdam, has conferred its freedom on me. To all this must be added countless individuals, scientific and personal friends, pupils, and others personally unknown to me, who have sent their congratulations in telegrams and in letters.

But this is not all. You desire to make my name the banner, as it were, of a magnificent institution which, founded by lovers of science of all nations, is to encourage and promote scientific inquiry in all countries. Science and art are, indeed, at the present time the only remaining bond of peace between civilised nations. Their ever-increasing development is a common aim of all; is effected by the common work of all, and for the common good of all. A great and a sacred work! The founders even wish to devote their gift to the promotion of those branches of science which all my life I have pursued, and thus bring me, with my shortcomings, before future generations almost as an exemplar of scientific investigation. This is the proudest honour which you could confer upon me, in so much as you thereby show that I possess your unqualified favourable opinion. But it would border on presumption were I to accept it without a quiet expectation on my part that the judges of future centuries will not be influenced by considerations of personal favour.

My personal appearance even, you have had represented in marble by a master of the first rank, so that I shall appear to the present and to future generations in a more ideal form; and another master of the etching needle has ensured that faithful portraits of me shall be distributed among my contemporaries.

I cannot fail to remember that all you have done is an expression of the sincerest and warmest goodwill on your part, and that I am most deeply indebted to you for it.

I must, however, be excused if the first effect of these abundant honours is rather surprising and confusing to me than intelligible. My own consciousness does not justify me in putting a measure of the value of what I have tried to do, which would leave such a balance in my favour as you have drawn. I know how simply everything I have done has been brought about; how scientific methods worked out by my predecessors have naturally led to certain results, and how frequently a fortunate circumstance or a lucky accident has helped me. But the chief difference is this —that which I have seen slowly growing from small beginnings through months and years of toilsome and tentative work, all that suddenly starts before you like Pallas fully equipped from the head of Jupiter. A feeling of surprise has entered into your estimate, but not into mine. At times, and perhaps even frequently, my own estimate may possibly have been unduly lowered by the fatigue of the work, and by vexation about all kinds of futile steps which I had taken. My colleagues, as well as the public at large, estimate a scientific or artistic work according to the utility, the instruction, or the pleasure which it has afforded. An author is usually disposed to base his estimate on the labour it has cost him, and it is but seldom that both kinds of judgment agree. It can, on the other hand, be seen from incidental expressions of some of the most celebrated men, especially of artists, that they lay but small weight on productions which seem to us inimitable, compared with others which have been difficult, and yet which appear to readers and observers as much less successful. I need only mention Goethe, who once stated to Eckermann that he did not estimate his poetical works so highly as what he had done in the theory of colours.

The same may have happened to me, though in a more modest degree, if I may accept your assurances and those of the authors of the addresses which have reached me. Permit me, therefore, to give you a short account of the manner in which I have been led to the special direction of my work.

In my first seven years I was a delicate boy, for long confined to my room, and often even to bed; but, nevertheless, I had a strong inclination towards occupation and mental activity. My parents busied themselves a good deal with me; picture books and games.

especially with wooden blocks, filled up the rest of the time. Reading came pretty early, which, of course, greatly increased the range of my occupations. But a defect of my mental organisation showed itself almost as early, in that I had a bad memory for disconnected things. The first indication of this I consider to be the difficulty I had in distinguishing between left and right; afterwards, when at school I began with languages, I had greater difficulties than others in learning words, irregular grammatical forms, and peculiar terms of expression. History as then taught to us I could scarcely master. To learn prose by heart was martyrdom. This defect has, of course, only increased, and is a vexation of my mature age.

But when I possessed small mnemotechnical methods, or merely such as are afforded by the metre and rhyme of poetry, learning by heart, and the retention of what I had learnt, went on better. I easily remembered poems by great authors, but by no means so easily the somewhat artificial verses of authors of the second rank. I think that is probably due to the natural flow of thought in good poems, and I am inclined to think that in this connection is to be found an essential basis of æsthetic beauty. In the higher classes of the Gymnasium I could repeat some books of the Odyssey, a considerable number of the odes of Horace, and large stores of German poetry. In other directions I was just in the position of our older ancestors, who were not able to write, and hence expressed their laws and their history in verse, so as to learn them by heart.

That which a man does easily he usually does willingly; hence I was first of all a great admirer and lover of poetry. This inclination was encouraged by my father, who, while he had a strict sense of duty, was also of an enthusiastic disposition, impassioned for poetry, and particularly for the classic period of German Literature. He taught German in the upper classes of the Gymnasium, and read Homer with us. Under his guidance we did, alternately, themes in German prose and metrical exercises—poems as we called them. But even if most of us remained indifferent poets, we learned better in this way, than in any other I know of, how to express what we had to say in the most varied manner.

But the most perfect mnemotechnical help is a knowledge of the laws of phenomena. This I first got to know in geometry. From the time of my childish playing with wooden blocks, the relations of special proportions to each other were well known to me from actual perception. What sort of figures were produced when bodies of regular shape were laid against each other I knew well without much consideration. When I began the scientific study of geometry, all the facts which I had to learn were perfectly well known and familiar to me, much to the astonishment of my teachers. So far as I recollect, that came out incidentally in the elementary school attached to the Potsdam Training College, which I attended up to my eighth year. Strict scientific methods, on the contrary, were new to me, and with their help I saw the difficulties disappear which had hindered me in other regions.

One thing was wanting in geometry; it dealt exclusively with abstract forms of space, and I delighted in complete reality. As I became bigger and stronger I went about with my father and my schoolfellows a great deal in the neighbourhood of my native town, Potsdam, and I acquired a great love of Nature. This is perhaps the reason why the first fragments of physics which I learned in the Gymnasium engrossed me much more closely than purely geometrical and algebraical studies. Here there was a copious and multifarious region, with the mighty fulness of Nature, to be brought under the dominion of a mentally apprehended law. And, in fact, that which first fascinated me was the intellectual mastery over Nature, which at first confronts us as so unfamiliar, by the logical force of law. But this, of course, soon led to the recognition that knowledge of natural processes was the magical key which places ascendency over Nature in the hands of its possessor. In this order of ideas I felt myself at home.

I plunged then with great zeal and pleasure into the study of all the books on physics I found in my father's library. They were very old-fashioned; phlogiston still held sway, and galvanism had not grown beyond the voltaic pile. A young friend and myself tried, with our small means, all sorts of experiments about which we had read. The action of acids on our mothers' stores of linen we investigated thoroughly; we had otherwise but little success. Most successful was, perhaps, the construction of optical instruments by means of spectacle glasses, which were to be had in Potsdam, and a small botanical lens belonging to my father. The limitation of our means had at that time the value that I was compelled always to vary in all possible ways my plans for experiments, until I got them in a form in which I could carry them out. I must confess that many a time when the class was reading Cicero or Virgil, both of which I found very tedious, I was calculating under the desk the path of rays in a telescope, and I discovered, even at that time, some optical theorems, not ordinarily met with in text-books, but which I afterwards found useful in the construction of the ophthalmoscope.

Thus it happened that I entered upon that special line of study to which I have subsequently adhered, and which, in the conditions I have mentioned, grew into an absorbing impulse, amounting even to a passion. This impulse to dominate the actual world by acquiring an understanding of it, or what, I think, is only another expression for the same thing, to discover the causal connection of phenomena, has guided me through my whole life, and the strength of this impulse is possibly the reason why I found no satisfaction in apparent solutions of problems so long as I felt there were still obscure points in them.

And now I was to go to the university. Physics was at that time looked upon as an art by which a living could not be made. My parents were compelled to be very economical, and my father explained to me that he knew of no other way of helping me to the study of Physics, than by taking up the study of medicine into the bargain. I was by no means averse from the study of living Nature, and assented to this without much difficulty. Moreover, the only influential person in our family had been a medical man, the late Surgeon-General Mursinna; and this relationship was a recommendation in my favour among other applicants for admission to our Army Medical School, the Friedrich Wilhelms Institut, which very materially helped the poorer students in passing through their medical course.

In this study I came at once under the influence of a profound teacher—Johannes Müller; he who at the same time introduced E. Du Bois Reymond, E. Brücke, C. Ludwig, and Virchow to the study of anatomy and physiology. As respects the critical questions about the nature of life, Müller still struggled between the older—essentially the metaphysical—view and the naturalistic one, which was then being developed; but the conviction that nothing could replace the knowledge of facts forced itself upon him with increasing certainty, and it may be that his influence over his students was the greater because he still so struggled.

Young people are ready at once to attack the deepest problems, and thus I attacked the perplexing question of the nature of the vital force. Most physiologists had at that time adopted G. E. Stahl's way out of the difficulty, that while it is the physical and chemical forces of the organs and substances of the living body which act on it, there is an indwelling vital soul or vital force which could bind and loose the activity of these forces; that after death the free action of these forces produces decomposition, while during life their action is continually being controlled by the soul of life. I had a misgiving that there was something against nature in this explanation; but it took me a good deal of trouble to state my misgiving in the form of a definite question. I found ultimately, in the latter years of my career as a student, that

Stahl's theory ascribed to every living body the nature of a *perpetuum mobile.* I was tolerably well acquainted with the controversies on this latter subject. In my school days I had heard it discussed by my father and our mathematical teachers, and while still a pupil of the Friedrich Wilhelms Institut I had helped in the library, and in my spare moments had looked through the works of Daniell, Bernouilli, D'Alembert, and other mathematicians of the last century. I thus came upon the question, 'What relations must exist between the various kinds of natural forces for a perpetual motion to be possible?' and the further one, 'Do those relations actually exist?' In my essay, 'On the Conservation of Force,' my aim was merely to give a critical investigation and arrangement of the facts for the benefit of physiologists.

I should have been quite prepared if the experts had ultimately said, 'We know all that. What is this young doctor thinking about, in considering himself called upon to explain it all to us so fully?' But, to my astonishment, the physical authorities with whom I came in contact took up the matter quite differently. They were inclined to deny the correctness of the law, and in the eager contest in which they were engaged against Hegel's Natural Philosophy were disposed to declare my essay to be a fantastical speculation. Jacobi, the mathematician, who recognised the connection of my line of thought with that of the mathematicians of the last century, was the only one who took an interest in my attempt, and protected me from being misconceived. On the other hand, I met with enthusiastic applause and practical help from my younger friends, and especially from E. Du Bois Reymond. These, then, soon brought over to my side the members of the recently formed Physical Society of Berlin. About Joule's researches on the same subject I knew at that time but little, and nothing at all of those of Robert Mayer.

Connected with this were a few smaller experimental researches on putrefaction and fermentation, in which I was able to furnish a proof, in opposition to Liebig's contention, that both were by no means purely chemical decompositions, spontaneously occurring, or brought about by the aid of the atmospheric oxygen; that alcoholic fermentation more especially was bound up with the presence of yeast spores which are only formed by reproduction. There was, further, my work on metabolism in muscular action, which afterwards was connected with that on the development of heat in muscular action; these being processes which were to be expected from the law of the conservation of force.

These researches were sufficient to direct upon me the attention of Johannes Müller as well as of the

Prussian Ministry of Instruction, and to lead to my being called to Berlin as Brücke's successor, and immediately thereupon to the University of Königsberg. The Army medical authorities, with thankworthy liberality, very readily agreed to relieve me from the obligation to further military service, and thus made it possible for me to take up a scientific position.

In Königsberg I had to lecture on general pathology and physiology. A university professor undergoes a very valuable training in being compelled to lecture every year, on the whole range of his science, in such a manner that he convinces and satisfies the intelligent among his hearers—the leading men of the next generation. This necessity yielded me, first of all, two valuable results.

For in preparing my course of lectures, I hit directly on the possibility of the ophthalmoscope, and then on the plan of measuring the rate of propagation of excitation in the nerves.

The ophthalmoscope is, perhaps, the most popular of my scientific performances, but I have already related to the oculists how luck really played a comparatively more important part than my own merit. I had to explain to my hearers Brücke's theory of ocular illumination. In this, Brücke was actually within a hair's breadth of the invention of the ophthalmoscope. He had merely neglected to put the question, To what optical image do the rays belong, which come from the illuminated eye? For the purpose he then had in view it was not necessary to propound this question. If he had put it, he was quite the man to answer it as quickly as I could, and the plan of the ophthalmoscope would have been given. I turned the problem about in various ways, to see how I could best explain it to my hearers, and I thereby hit upon the question I have mentioned. I knew well, from my medical studies, the difficulties which oculists had about the conditions then comprised under the name of Amaurosis, and I at once set about constructing the instrument by means of spectacle glasses and the glass used for microscope purposes. The instrument was at first difficult to use, and without an assured theoretical conviction that it must work, I might, perhaps, not have persevered. But in about a week I had the great joy of being the first who saw clearly before him a living human retina.

The construction of the ophthalmoscope had a very decisive influence on my position in the eyes of the world. From this time forward I met with the most willing recognition and readiness to meet my wishes on the part of the authorities and of my colleagues, so that for the future I was able to pursue far more freely the secret impulses of my desire for knowledge. I must, however, say that I ascribed my success in great measure to the circumstance that, possessing some geometrical capacity, and equipped with a knowledge of physics, I had, by good fortune, been thrown among medical men, where I found in physiology a virgin soil of great fertility; while, on the other hand, I was led by the consideration of the vital processes to questions and points of view which are usually foreign to pure mathematicians and physicists. Up to that time I had only been able to compare my mathematical abilities with those of my fellow-pupils and of my medical colleagues; that I was for the most part superior to them in this respect did not, perhaps, say very much. Moreover, mathematics was always regarded in the school as a branch of secondary rank. In Latin composition, on the contrary, which then decided the palm of victory, more than half my fellow-pupils were ahead of me.

In my own consciousness, my researches were simple logical applications of the experimental and mathematical methods developed in science, which by slight modifications could be easily adapted to the particular object in view. My colleagues and friends, who, like myself, had devoted themselves to the physical aspect of physiology, furnished results no less surprising.

But in the course of time matters could not remain in that stage. Problems which might be solved by known methods I had gradually to hand over to the pupils in my laboratory, and for my own part turn to more difficult researches, where success was uncertain, where general methods left the investigator in the lurch, or where the method itself had to be worked out.

In those regions also which come nearer the boundaries of our knowledge I have succeeded in many things experimental and mechanical— I do not know if I may add philosophical. In respect of the former, like any one who has attacked many experimental problems, I had become a person of experience, who was acquainted with many plans and devices, and I had changed my youthful habit of considering things geometrically into a kind of mechanical mode of view. I felt, intuitively as it were, how strains and stresses were distributed in any mechanical arrangement, a faculty also met with in experienced mechanicians and machine constructors. But I had the advantage over them of being able to make complicated and specially important relations perspicuous, by means of theoretical analysis.

I have also been in a position to solve several mathematical physical problems, and some, indeed, on which the great mathematicians, since the time of Euler, had in vain occupied themselves; for example, questions as to vortex motion and the discontinuity of motion in liquids, the question as to the motion of sound at the open ends of organ pipes, &c. &c. But the pride which I might have felt about the final result in these cases was considerably lowered by my consciousness that I had only succeeded in solving such problems after many devious ways, by the gradually increasing generalisation of favourable examples, and by a series of fortunate guesses. I had to compare myself with an Alpine climber, who, not knowing the way, ascends slowly and with toil, and is often compelled to retrace his steps because his progress is stopped; sometimes by reasoning, and sometimes by accident, he hits upon traces of a fresh path, which again leads him a little further; and finally, when he has reached the goal, he finds to his annoyance a royal road on which he might have ridden up if he had been clever enough to find the right starting-point at the outset. In my memoirs I have, of course, not given the reader an account of my wanderings, but I have described the beaten path on which he can now reach the summit without trouble.

There are many people of narrow views, who greatly admire themselves, if once in a way, they have had a happy idea, or believe they have had one. An investigator, or an artist, who is continually having a great number of happy ideas, is undoubtedly a privileged being, and is recognised as a benefactor of humanity.

But who can count or measure such mental flashes? Who can follow the hidden tracts by which conceptions are connected?

> That which man had never known,
> Or had not thought out,
> Through the labyrinth of mind
> Wanders in the night.

I must say that those regions, in which we have not to rely on lucky accidents and ideas, have always been most agreeable to me, as fields of work.

But, as I have often been in the unpleasant position of having to wait for lucky ideas, I have had some experience as to when and where they came to me, which will perhaps be useful to others. They often steal into the line of thought without their importance being at first understood; then afterwards some accidental circumstance shows how and under what conditions they have originated; they are present, otherwise, without our knowing whence they came. In other cases they occur suddenly, without exertion, like an inspiration. As far as my experience goes, they never came at the desk or to a tired brain. I have always so turned my problem about in all directions that I could see in my mind its turns and complications, and run through them freely without writing them down. But to reach that stage was not usually possible without long preliminary work. Then, after the fatigue from this had passed away, an hour of perfect bodily repose and quiet comfort was necessary before the good ideas came. They often came actually in the morning on waking, as expressed in Goethe's words which I have quoted, and as Gauss also has remarked.[1] But, as I have stated in Heidelberg, they were usually apt to come when comfortably ascending woody hills in sunny weather. The smallest quantity of alcoholic drink seemed to frighten them away.

Such moments of fruitful thought were indeed very delightful, but not so the reverse, when the redeeming ideas did not come. For weeks or months I was gnawing at such a question until in my mind I was

> Like to a beast upon a barren heath
> Dragged in a circle by an evil spirit,
> While all around are pleasant pastures green.

And, lastly, it was often a sharp attack of headache which released me from this strain, and set me free for other interests.

I have entered upon still another region to which I was led by investigation on perception and observation of the senses, namely, the theory of cognition. Just as a physicist has to examine the telescope and galvanometer with which he is working; has to get a clear conception of what he can attain with them, and how

[1] Gauss, *Werke*, vol. v. p. 609. 'The law of induction discovered Jan. 23, 1835, at 7 A.M., before rising.'

they may deceive him; so, too, it seemed to me necessary to investigate likewise the capabilities of our power of thought. Here, also, we were concerned only with a series of questions of fact about which definite answers could and must be given. We have distinct impressions of the senses, in consequence of which we know how to act. The success of the action usually agrees with that which was to have been anticipated, but sometimes also not, in what are called subjective impressions. These are all objective facts, the laws regulating which it will be possible to find. My principal result was that the impressions of the senses are only signs for the constitution of the external world, the interpretation of which must be learned by experience. The interest for questions of the theory of cognition, had been implanted in me in my youth, when I had often heard my father, who had retained a strong impression from Fichter's idealism, dispute with his colleagues who believed in Kant or Hegel. Hitherto I have had but little reason to be proud about those investigations. For each one in my favour, I have had about ten opponents; and I have in particular aroused all the metaphysicians, even the materialistic ones, and all people of hidden metaphysical tendencies. But the addresses of the last few days have revealed a host of friends whom as yet I did not know; so that in this respect also I am indebted to this festivity for pleasure and for fresh hope. Philosophy, it is true, has been for nearly three thousand years the battle-ground for the most violent differences of opinion, and it is not to be expected that these can be settled in the course of a single life.

I have wished to explain to you how the history of my scientific endeavours and successes, so far as they go, appears when looked at from my own point of view, and you will perhaps understand that I am surprised at the universal profusion of praise which you have poured out upon me. My successes have had primarily this value for my own estimate of myself, that they furnished a standard of what I might further attempt; but they have not, I hope, led me to self-admiration. I have often enough seen how injurious an exaggerated sense of self-importance may be for a scholar, and hence I have always taken great care not to fall a prey to this enemy. I well knew that a rigid self-criticism of my own work and my own capabilities was the protection and palladium against this fate. But it is only needful to keep the eyes open for what others can do, and what one cannot do oneself, to find there is no great danger; and, as regards my own work, I do not think I have ever corrected the last proof of a memoir without finding in the course of twenty-four hours a few points which I could have done better or more carefully.

As regards the thanks which you consider you owe me, I should be unjust if I said that the good of humanity appeared to me, from the outset, as the conscious object of my labours. It was, in fact, the special form of my desire for knowledge which impelled me and determined me, to employ in scientific research all the time which was not required by my official duties and by the care for my family. These two restrictions did not, indeed, require any essential deviation from the aims I was striving for. My office required me to make myself capable of delivering lectures in the University; my family, that I should establish and maintain my reputation as an investigator. The State, which provided my maintenance, scientific appliances, and a great share of my free time, had, in my opinion, acquired thereby the right that I should communicate faithfully and completely to my fellow-citizens, and in a suitable form, that which I had discovered by its help.

The writing out of scientific investigations is usually a troublesome affair; at any rate it has been so to me. Many parts of my memoirs I have rewritten five or six times, and have changed the order about until I was fairly satisfied. But the author has a great advantage in such a careful wording of his work. It compels him to make the severest criticism of each sentence and each conclusion, more thoroughly even than the lectures at the University which I have mentioned. I have never considered an investigation finished until it was formulated in writing, completely and without any logical deficiencies.

Those among my friends who were most conversant with the matter represented to my mind, my conscience as it were. I asked myself whether they would approve of it. They hovered before me as the embodiment of the scientific spirit of an ideal humanity, and furnished me with a standard.

In the first half of my life, when I had still to work for my external position, I will not say that, along with a desire for knowledge and a feeling of duty as servant of the State, higher ethical motives were not also at work; it was, however, in any case difficult to be certain of the reality of their existence so long as selfish motives were still existent. This is, perhaps, the case with all investigators. But afterwards, when an assured position has been attained, when those who have no inner impulse towards science may quite cease their labours, a higher conception of their relation to humanity does influence those who continue to work. They gradually learn from their own experience how the thoughts which they have uttered, whether through literature or through oral instruction, continue to act on their fellow-men, and possess, as it were, an independent life; how these thoughts, further worked out by their pupils, acquire a deeper significance and a more definite form, and, reacting on their originators, furnish them with fresh instruction. The ideas of an individual, which he himself has conceived, are of course more closely connected with his mental field of view than extraneous ones, and he feels more encouragement and satisfaction when he sees the latter more abundantly developed than the former. A kind of parental affection for such a mental child ultimately springs up, which leads him to care and to struggle for the furtherance of his mental offspring as he does for his real children.

But, at the same time, the whole intellectual world of civilised humanity presents itself to him as a continuous and spontaneously developing whole, the duration of which seems infinite as compared with that of a single individual. With his small contributions to the building up of science, he sees that he is in the service of something everlastingly sacred, with which he is connected by close bands of affection. His work thereby appears to him more sanctified. Anyone can, perhaps, apprehend this theoretically, but actual personal experience is doubtless necessary to develop this idea into a strong feeling.

The world, which is not apt to believe in ideal motives, calls this feeling love of fame. But there is a decisive criterion by which both kinds of sentiment can be discriminated. Ask the question if it is the same thing to you whether the results of investigation which you have obtained are recognised as belonging to you or not when there are no considerations of external advantage bound up with the answer to this question. The reply to it is easiest in the case of chiefs of laboratories. The teacher must usually furnish the fundamental idea of the research as well as a number of proposals for overcoming experimental difficulties, in which more or less ingenuity comes into play. All this passes as the work of the student, and ultimately appears in his name when the research is finished. Who can afterwards decide what one or the other has done? And how many teachers are there not who in this respect are devoid of any jealousy?

Thus, gentlemen, I have been in the happy position that, in freely following my own inclination, I have been led to researches for which you praise me, as having been useful and instructive. I am extremely fortunate that I am praised and honoured by my contemporaries, in so high a degree, for a course of work which is to me the most interesting I could pursue. But my contemporaries have afforded me great and essential help. Apart from the care for my own existence and that of my family, of which they have relieved me, and apart from the external means with which they have provided me, I have found in them a standard of the intellectual capacity of man; and by their sympathy for my work they have evoked in me a vivid conception of the universal mental life of humanity which has enabled me to see the value of my own researches in a higher light. In these circumstances, I can only regard as a free gift the thanks which you desire to accord to me, given unconditionally and without counting on any return.